# UNDERSTANDING CATCHMENT PROCESSES AND HYDROLOGICAL MODELLING IN THE ABAY/UPPER BLUE NILE BASIN, ETHIOPIA

# UNDERSTANDING CATCHMENT PROCESSES AND HYDROLOGICAL MODELLING IN THE ABAY/UPPER BLUE NILE BASIN, ETHIOPIA

## DISSERTATION

Submitted in fulfilment of the requirements of
the Board for Doctorates of Delft University of Technology and of
the Academic Board of the UNESCO-IHE Institute for Water Education
for the Degree of DOCTOR
to be defended in public
on Tuesday, 03 February 2015 at 15:00 hrs
in Delft, The Netherlands

by

Sirak Tekleab Gebrekristos

Master of Science Degree in Hydrology and Water Resources Management, Arba Minch
University, Ethiopia
born in Addis Ababa, Ethiopia

CRC Press
Taylor & Francis Group
Boca Raton  London  New York

CRC Press is an imprint of the
Taylor & Francis Group, an **informa** business
A BALKEMA BOOK

This dissertation has been approved by the promoters:

Prof. dr. S. Uhlenbrook
Prof. dr. ir. H.H.G. Savenije

Composition of Doctoral Awarding Committee:

| | |
|---|---|
| Chairman | Rector Magnificus, Delft University of Technology |
| Vice-chairman | Rector UNESCO-IHE |
| Prof. dr. S. Uhlenbrook | UNESCO-IHE/Delft University of Technology, promoter |
| Prof. dr. ir. H.H.G. Savenije | Delft University of Technology/UNESCO-IHE, promoter |
| Prof. dr. M.E. McClain | UNESCO-IHE/Delft University of Technology |
| Prof. dr. ir. N.C. van de Giesen | Delft University of Technology |
| Prof. dr. ir. P. van der Zaag | UNESCO-IHE/Delft University of Technology |
| Prof. dr. Y.A. Mohamed | HRC, Sudan/UNESCO-IHE |
| Prof. dr. W. G. M. Bastiaanssen | Delft University of Technology/UNESCO-IHE (reserve member) |

The research reported in this dissertation has been funded by the Netherlands Organisation for Scientific Research (WOTRO).

First issued in hardback 2018

*CRC Press/Balkema is an imprint of the Taylor & Francis Group, an informa business*

Published by:
CRC Press/Balkema
PO Box 11320, 2301 EH Leiden, The Netherlands
e-mail: Pub.NL@taylorandfrancis.com
www.crcpress.com – www.taylorandfrancis.com

ISBN 13: 978-1-138-37330-3 (hbk)
ISBN 13: 978-1-138-02792-3 (pbk)

# ACKNOWLEDGEMENTS

First of all, I would like to thank God for his never-ending love, care, and giving me stamina to accomplish this research well in a given period of time. This research would not have been possible without proper guidance, support, and encouragements from different people and organisations.

My heartfelt gratitude goes to my promoter Prof. Dr. Stefan Uhlenbrook, who has given me a chance to do my PhD under his guidance. Stefan, I appreciate your intellectual scientific capabilities and the timely response to all the academic and non-academic matters related to my research. I have learned a lot from you. You are always optimistic and see things in different angles. Your expertise knowledge as a modeller helped me a lot to understand how catchment hydrological studies work and particularly helped me to better understand data scarce environments.

I am also indebted to my second promoter Prof. Dr. Ir. Huub Savenije, who has contributed to my research in many ways. I well-regarded your broad knowledge related to the advances of hydrological science through scientific research and teaching. I enjoyed your advanced hydrological modelling course at Delft, University of Technology water resources section. I have got countless valuable concepts, ideas, and knowledge about hydrological science in general and hydrological modelling in particular, which helped me, a lot to apply the acquired knowledge for this thesis research.

I am grateful to my supervisors Dr. Yasir Abbas and Dr. Jochen Wenninger for their consistent guidance, critical comments and suggestions to bring the thesis to the current form. Dr. Yasir, I appreciated your knowledge of basin hydrology in general and the Blue Nile in particular, that helped me a lot to interpret the results in a scientific way. Dr. Jochen, your involvement throughout the research period was highly appreciated. You offered an introductory course on the isotope techniques for hydrological applications, field instrumentation, and laboratory analysis; this helped me a lot to gain the knowledge summarised in this thesis.

I would like to thank the Blue Nile hydro-solidarity research project director's Prof. Dr. ir. Pieter Van der Zaag (UNESCO-IHE) and Dr. Belay Simane (Addis Ababa University) for their roles in coordinating the research activities related to the academic, financial and administrative issues.

I am also thankful to the Dutch Foundation for the Advancement of Tropical Research (WOTRO), for financing all the costs related to this research throughout the research period. I would take this opportunity to thank Ms. Jolanda Boots PhD fellowship officer at UNESCO-IHE, Delft, who managed the administrative and financial issues related to my research work. I would like to thank my employer Hawassa University, Institute of Technology, Ethiopia, for granting me a study leave to execute this research work. My special thanks also go to my head of department Mr. Wossenu Lemma, who reduced my work load in order to focus on this research.

My sincere thanks also go to all the PhD researchers, namely Dr. Abonesh, Ermias, Ishraga, Hermen, Khalid, Dr. Melesse (post-doctoral researcher), Rahel, Reem, Twedros and Dr. Yasir Salih, who have been conducting their research in the Blue Nile hydro-solidarity project. I really appreciate the integrative research efforts we have made jointly, and the generation and sharing of multi-disciplinary knowledge about the basin. I extend my special thanks to all PhD research fellows and MSc participants at UNESCO-IHE, Delft. I am also grateful to Mr. Joost Hollander for translating the summary in to the Dutch language.

Special thanks go to my colleagues Akelilu Dinkeneh (Aki, you did a lot for me!), Adinew Berga (no words to thank you!), Eskinder, Fikadu, Kedir Adal, Dr. Ing. Mebruk, Dr. Ing. Netsanet, Selam Assefa, Sileshi, Dr. Solomon Seyoum at Unesco-ihe, Twedros Meles, Yared Ashenafi, Yared Abebe, and Zeleke for their support and encouragements throughout the research period. I am also grateful to my special friends Dr. Girma Yimer and Tizitaw Tefera who are always happy to discuss both academic and non-academic matters with me. Dr. Girma, you are really a good friend and I very much appreciate your kindness in helping others. Keep up your good work! Furthermore, I would like to thank colleagues at Hawassa University, particularly Dr. Awdenegest Moges, Mr. Alemayehu Muluneh and Mr. Getahun Alemu for sharing the same office at Hawassa College of Agriculture and many fruitful discussions during the tea breaks.

The field work activities would not have been possible without the support I got from the development agents, farmers and other individuals, namely Chekolech Mengiste, Derseh Gebeyehu, Lakachew Alemu, Limenew Mihrete, Mengistu Abate and Solomon Takele. I thank all of them for their kind cooperation and hard work. My sincere gratitude also goes to the Ministry of Water, Irrigation, and Energy and the National Meteorological Service Agency for providing the streamflow and weather data sets, respectively.

I owe many thanks to my parents Tekleab Gebrekristos and Azalech Ayele for their continuous encouragements throughout the research period. Abiy, Asres, Bethelhem, Dawit, Deneke, Kidist, Habtewold, Tsegaye (Chale), mother-in-law Worknesh, Woyeneshet and Yidenekachew; your significant contributions are highly appreciated and let to the successful completion of my study.

Last but not least, I would like to thank my wife Meseret Hailu for her consistent love, care and encouragements throughout the research period. My son Sofonias ('Sofi') missed me a lot and I missed you as well while I was in the Netherlands. Sofi, you are my real sources of inspiration. My little daughter Heran ('titiye'), you are so lucky that you were born during the final phase of my PhD study that I have more time for you during your development stage.

*This thesis is dedicated to the rural women in Ethiopia.*

Sirak Tekleab Gebrekristos

September, 2014, Hawassa, Ethiopia

# SUMMARY

The water resources in the Abay/Upper Blue Nile basin are the source of life for the several hundred million people living in the basin and further downstream. It provides more than 60% of the total Nile water. Intensive farming in unfavourable soils and slopes, overgrazing and soil erosion is among the major problems in the basin. Land degradation as a result of soil erosion decreases soil fertility in the upper catchments and simultaneously increases sedimentation in reservoirs and irrigation canals in downstream countries. Land degradation also affects basin hydrology and water resources availability. Therefore, sustainable water resources management in the basin is necessary that requires in-depth understanding of the basin hydrology. This could be achieved in one way through assessment of hydrological variability, investigating the two-way interactions between land use on the hydrologic responses, and detailed understanding of the rainfall runoff processes.

Although detailed knowledge of the basin hydrology is important both from scientific and operational perspectives, it is hindered by the scarce hydro-meteorological data. Besides, the space-time variations of rainfall and other meteorological parameters as well as physiographic characteristics are large and, consequently, the hydrological processes are quite complex in the basin. Furthermore, these processes have been rarely investigated before. Therefore, analysing the hydrological processes at various spatio-temporal scales has the first priority to be able to predict the impact of changes in the basin and guide sustainable water resources management. This thesis aims at characterising and quantifying catchment processes and modelling in the basin through intensive field measurements and a set of different modelling approaches that complement each other in the range of space and time scales. Different methods including long-term trend analysis, field data collection and combined stable isotopes and process based rainfall-runoff modelling were carried out.

The long-term trends (1954-2010) of rainfall, temperature and streamflow were analysed for three gauging stations in the basin. Mann-Kendall and Pettitt tests were used for the trend and detection of change point analysis, respectively. The results showed abrupt changes and related upward and downward shifts in temperature and streamflow time series, respectively. However, precipitation time series did not reveal any statistically significant trends at 5% significance level in mean annual and seasonal scales across the examined stations. Increasing trends in temperature at different weather stations for the mean annual, rainy, dry and short rainy seasons became apparent, e.g., the mean temperature at Bahir Dar in the Lake Tana sub basin has increased at a rate of 0.5, 0.3, and 0.6 °C/decade for the main rainy season (June to September), short rainy season (March to May) and the dry season (October to February), respectively.

To understand the large-scale hydrological dynamics of the basin, water balances have been computed for twenty selected catchments ranging from 200 to 173,686 km$^2$. A simple model based on Budyko's hypothesis following a top-down modelling approach was used to analyse the water balance on annual and monthly timescales. The results showed that at annual timescale all catchments cannot be represented by the same Budyko curve. Inter-annual variability of rainfall, land use/landcover, soil type, geology, and topography are most likely the reason for different catchment responses. The results of the annual water balance model were improved by reducing the timescale into monthly and by incorporating the monthly dynamics of the root zone soil moisture storage into the model. The monthly model produced better calibration and validation results of the streamflow dynamics for the majority of the catchments.

The results of the trend analyses and water balance computations showed considerable variations in flow regimes and hydrologic responses across the Abay/Upper Blue Nile catchments. To gain further insights into the reasons for the variability an in-depth investigation of the hydrological responses to land use/land cover changes was conducted for the meso-scale Jedeb catchment (296 km$^2$). The Jedeb catchment is characterised by intensive farming and considerable expansion of agricultural land during the last few decades, e.g. agricultural land has increased by 17% in the last 50 years.

Using statistical tests for daily data and monthly modelled data based on the Budyko approach the effect of land use change on the streamflow was quantified for the Jedeb catchment. The results reveal a significant change of the daily flow characteristics observed between 1973 and 2010. Peak flow is increased, i.e., the response has become flashier. There is a significant increase in the rise and fall rates of the hydrograph as well as in the number of low flow pulses. The discharge pulses show a declining duration with time, which is a sign of increased 'flashiness' of the catchment. The Budyko model results demonstrated a change in model parameters over time, which could also be attributed to a land use/land cover change. The model parameters representing soil moisture conditions indicated a gradual decreasing trend, implying reducing storage capacity, which can be attributed to increasing farming in the catchment accompanied by a relative decrease in natural land covers types including forest. The results of the monthly flow duration curve analysis indicated large changes of the flow regime over time. The high flows have increased by 45% between the 1990s and 2000s. Whereas low flows decreased by 85% between the 1970s and 2000s. These results are relevant to guide sustainable catchment management practices in the Jedeb catchment and in other similar catchments within the basin.

The characterisation of stable environmental isotopes to identify mean residence times and runoff components in the headwater catchments provided useful information for the management of the catchment. Both in Jedeb and the neighbouring meso-scale catchment

Chemoga, in-situ isotope samples of precipitation, spring water and streamflow were collected and analysed. The results show that the isotopic composition of precipitation exhibits marked seasonal variations, which suggests different sources of moisture generating the rainfall in the study area. The Atlantic Ocean, Congo basin, Upper White Nile and the Sudd swamps were identified as potential moisture source areas during the main rainy (summer) season, while, the Indian-Arabian, and Mediterranean Sea are the main moisture source areas during the short rainy season and the dry (winter) seasons. Results from the hydrograph separation at a seasonal time scale indicate the dominance of event water with an average of 71% and 64% of the total runoff during the wet season in the Chemoga and Jedeb catchment, respectively. The results further demonstrated that the mean residence times of the stream water are 4.1 and 6.0 months for the whole Chemoga and Jedeb catchments, respectively.

Finally, based on the field measurements, catchment response analysis and water balance modelling, a detailed process-based hydrological model were developed using the PCRaster software environment. The model was developed to study the hydrological processes of the two meso-scale catchments of Chemoga and Jedeb on a daily timescale. The measured forcing data, daily discharge series and various spatial data sets were used to develop the model at a grid size of 200*200 m$^2$. Three different model representations were employed to obtain an appropriate model structure. Model calibration and uncertainty assessment were implemented within the Generalized Likelihood Uncertainty Estimation (GLUE) framework. During model calibration, parameters were conditioned using the discharge data alone as well as stable environmental isotope information as a fraction of new water and old water percentages. The results from the different model representations were evaluated in terms of performance measures, parameter identifiability and reduced predictive uncertainty.

The model results clearly demonstrated that parameters appeared to be better identifiable and have a reduced model predictive uncertainty when using stable isotope information in addition to runoff measurements. The stable isotope data provided additional information about flow pathways and runoff components in the two catchments and, consequently, supported the process-based modelling. It has been found that the saturation excess overland flow is most likely the dominant runoff generation process during rainfall/runoff events in both catchments, which is in line with field observations. The model investigations demonstrated that the two catchments cannot be modelled equally well with the same model structure. This is attributed to differences in the rainfall-runoff processes caused by different size wetlands in each catchment. It is therefore concluded that a single model structure in a lumped way for the entire Abay/Upper Blue Nile cannot represent all dominant hydrological processes of the sub-catchments. Thus, semi-distributed modelling with distributed forcing inputs which account for the individual runoff processes assigned to different landscape features e.g. wetlands, hillslopes, and plateau is essential.

Given the results obtained by detailed hydrological measurements at two meso-scale sub-catchments, the hydrologic responses to land use/land cover change, the long-term trend analysis of hydro-meteorological parameters, the large scale Budyko modelling and, finally, the detailed conceptual distributed modelling, this study has provided in-depth insights and a better understanding of the hydrological processes within the Upper Blue Nile Basin. This is important for the management and sustainable development of the Blue Nile water resources as well as for future research in the basin.

## List of Symbols

| Symbol | Description | Dimension |
|---|---|---|
| a | Rating curve constant | $(L^{3-b}\,T^{-1})$ |
| b | Rating curve constant | ( - ) |
| A | Amplitude of predicted $\delta^{18}O$ | (‰) |
| h | Water level | (L) |
| $h_0$ | Water level for zero discharge | (L) |
| i | Slope | ( - ) |
| $\beta$ | Shape factor for runoff generation | ( - ) |
| $C_f$ | Angular frequency constant | $(rad^{-1})$ |
| C | Uptake of moisture from saturated zone to root zone storage | $(L\,T^{-1})$ |
| $C_T$ | Streamflow isotope concentration | (‰) |
| $C_E$ | Event water isotope concentration | (‰) |
| $C_{pe}$ | Pre-event water isotope concentration | (‰) |
| d | Time scale for slow flow storage | $(T^{-1})$ |
| $\dfrac{dS}{dt}$ | Change in water storage of the catchment over time | $(L\,T^{-1})$ |
| $E_i$ | Evaporation from intercepted surface | $(L\,T^{-1})$ |
| $E_p$ | Potential evaporation | $(L\,T^{-1})$ |
| E | Evaporation | $(L\,T^{-1})$ |
| $E_a$ | Actual evaporation | $(L\,T^{-1})$ |
| $H_0$ | Null hypothesis | ( - ) |
| $H_a$ | Alternative hypothesis | (-) |
| $K_T$ | Test statistic for the Pettitt test | (-) |
| $\phi$ | Aridity index | (-) |
| $\Phi$ | Phase lag | (rad) |
| $K_f$ | Time scale constant for fast reservoir | $(T^{-1})$ |
| $K_s$ | Time scale constant for slow reservoir | $(T^{-1})$ |
| $I_{th}$ | Maximum threshold depth for interception reservoir | $(L\,T^{-1})$ |
| $L_p$ | Fraction constraining potential transpiration | (-) |
| n | Manning roughness coefficient | $(L^{-1/3}\,T)$ |
| Q | Discharge | $(L\,T^{-1},\,L^3\,T^{-1})$ |
| $Q_b$ | Base flow discharge | $(L\,T^{-1},\,L^3\,T^{-1})$ |
| $Q_d$ | Direct runoff | $(L\,T^{-1},\,L^3\,T^{-1})$ |

| $Q_f$ | Fast discharge | $(L\,T^{-1}, L^3\,T^{-1})$ |
|---|---|---|
| $Q_{OF}$ | Fast overland flow | $(L\,T^{-1}, L^3\,T^{-1})$ |
| $Q_{Pe}$ | Pre event discharge | $(L\,T^{-1}, L^3\,T^{-1})$ |
| $Q_E$ | Event discharge | $(L\,T^{-1}, L^3\,T^{-1})$ |
| $Q_T$ | Total discharge | $(L\,T^{-1}, L^3\,T^{-1})$ |
| $\Theta$ | Parameter sets | $(-)$ |
| $\alpha_1$ | Rainfall retention parameter | $(-)$ |
| $\alpha_2$ | Evaporation coefficient | $(-)$ |
| $\alpha$ | Runoff partitioning coefficient | $(-)$ |
| $p$ | Probability value | $(-)$ |
| $P$ | Precipitation | $(L\,T^{-1})$ |
| $P_c$ | Maximum percolation rate | $(L\,T^{-1})$ |
| $P_e$ | Effective rainfall | $(L\,T^{-1})$ |
| $P_r$ | Preferential recharge | $(L\,T^{-1})$ |
| $r_1$ | Auto correlation coefficient | $(-)$ |
| $R$ | Recharge | $(L\,T^{-1})$ |
| $R_c$ | Runoff coefficient | $(-)$ |
| $R_s$ | Flux entering into fast storage reservoir | $(L\,T^{-1})$ |
| $R_u$ | Infiltrating flux into root zone soil moisture storage | $(L\,T^{-1})$ |
| $R_{sp}$ | Spearman rank correlation coefficient | $(-)$ |
| $\delta^{18}O$ | Oxygen isotope concentration | $(‰)$ |
| $\delta^2H$ | Hydrogen isotope concentration | $(‰)$ |
| $S$ | Mann-Kendall test statistics | $(-)$ |
| $S_{max}$ | Maximum root zone soil moisture storage capacity | $(L)$ |
| $S_f$ | Fast discharge storage reservoir | $(L)$ |
| $S_u$ | Root zone soil moisture storage | $(L)$ |
| $S_s$ | Slow flow storage reservoir | $(L)$ |
| $S_w$ | Wetland storage | $(L)$ |
| $S_{f,\,th}$ | Maximum threshold depth for fast storage reservoir | $(L)$ |
| $S_{s,\,th}$ | Maximum threshold depth for slow storage reservoir | $(L)$ |
| $S_{w,\,th}$ | Maximum threshold depth for wetland storage | $(L)$ |
| $T$ | Air temperature | $(°C)$ |
| $T_{rs}$ | Residence time | $(T)$ |
| $T_p$ | Potential transpiration | $(L\,T^{-1})$ |

| | | |
|---|---|---|
| $w$ | Coefficient representing integrated effects of catchment characteristics | (-) |
| W | Total uncertainty | (‰) |
| $X_t$ | Original hydro-meteorological series for Mann-Kendall test | (L T$^{-1}$, L$^3$ T$^{-1}$, °C) |
| $Y_t$ | De-trended hydro-meteorological series for Mann-Kendall test | (L T$^{-1}$, L$^3$ T$^{-1}$, °C) |
| $Y_2$ | New hydro-meteorological series for Mann-Kendall test | (L T$^{-1}$, L$^3$ T$^{-1}$, °C) |
| Z | Standard normal variate | (-) |

# List of Acronyms

| | |
|---|---|
| ANN | Artificial Neural Network |
| a.s.l | Above sea level |
| BCEOM | Le Bureau Central d'Etudes pour les Equipements 'Outre-Mer |
| CGIR-CSI | The Consultative Group on International Agricultural Research Consortium for Spatial Information |
| DEM | Digital Elevation Model |
| ENMSA | Ethiopian National Meteorological Service Agency |
| EEPCO | Ethiopian Electric Power Corporation |
| ENSO | El Niño–Southern Oscillation |
| ENTRO | Easter Nile Technical and Regional Office |
| ERT | Electrical Resistivity Tomography |
| FAO | Food and Agriculture Organization |
| FDREMW | Federal Democratic Republic of Ethiopia Ministry of Water Resources |
| GCM | Global Circulation Model |
| GLUE | Generalized Likelihood Uncertainty Estimation |
| GMWL | Global Meteoric Water Line |
| GERD | Grand Ethiopian Renaissance Dam |
| GWS | Groundwater Storage |
| ha | Hectare |
| HAND | Height Above the Nearest Drainage |
| HBV | Hydrologiska Byrans Vattenbalansavdelning |
| HRC | Hydraulics Research Center |
| HYSPLIT | Hybrid Single Particle Lagrangian Integrated Trajectory |

| | |
|---|---|
| IAHS | International Association of Hydrological Science |
| ITCZ | Inter Tropics Convergent Zone |
| LMWL | Local Meteoric Water Line |
| LULC | Land use Land cover |
| MW | Mega Watt |
| NBI | Nile basin Initiative |
| NOAA | National Oceanic and Atmospheric Administration |
| PUB | Prediction in Ungauged Basin |
| RSA | Regional Sensitivity Analysis |
| RZS | Root Zone Storage |
| SCRP | Soil Conservation Research Project |
| SENSE | Socio-Economic and Natural Sciences for the Environment |
| SHE | System Hydrologique European |
| SRTM | Shuttle radar topographic Mission |
| SWAT | Soil and Water Assessment Tool |
| TFPW | Trend Free Pre Whitening |
| UNESCO | United Nation Educational Scientific and Cultural Organization |
| USSR | Union of Soviet Socialist Republics |
| VSMOW | Vienna Standard Mean Ocean Water |
| WOTRO | The Dutch Foundation for the Advancement of Tropical Research |

# TABLE OF CONTENTS

# CHAPTER 1

## INTRODUCTION

### 1.1 Background

The growing demands of water use and tension in shared water resources like the Nile basin require scientific research on climatic and hydrological processes to support sustainable land and water development. Multi-disciplinary research is essential to understand the societal, economic, political and environmental perspectives of the shared water resources. The Abay/Upper Blue Nile River is one of the main tributaries of the Nile River originating from the Ethiopian highlands flowing to Sudan, where it meets the White Nile at Khartoum to form the Main Nile travelling north to Egypt and finally into the Mediterranean Sea. The basin is a typical example depicting interdependencies of upstream and downstream water users through the water and sediment fluxes linking the Eastern Nile countries Ethiopia, Sudan and Egypt. The Abay/Blue Nile contributes the major part of the Nile water with more than 60% of the total discharge. The flow is seasonal, depending on the main rainy period from June to September. Hundreds of millions of people depend on Blue Nile water in the upstream part Ethiopia as well as in the downstream Sudan and Egypt. Rain-fed and irrigated agriculture provide the main source of livelihoods for majority of the people in the basin.

The population in the basin is facing many problems. Poverty and limited development is common in many parts of the basin. Climatic and hydrological extremes such as floods and droughts hit the basin population severly and regularly. The soil erosion upstream contributes to lower rate of food production due to lower soil fertility in the rain-fed catchments. Consequently, the operation and maintenance costs downstream with regard to reservoir and canal sedimentation management are escalating. The problems of scarce hydro-climatic data and limited hydrological studies are also well-known in the basin. Furthermore, the current rapidly growing economies, population growth and urbanization urge for secured food supplies and sustainable water resources development in the basin.

Thus, research on hydrology should be the first priority and essential to inform better water management policies and strategies in the basin. This thesis provides an in-depth analysis to better understand and characterise the climate and hydrology of the basin at various spatial and temporal scales. To this end, understanding hydrological variability and catchment hydrological processes using hydrological models have become of paramount importance for investigating impacts of floods and droughts, land use and climatic changes, water quality and quantity in the basin.

## 1.2 Understanding hydrological processes

An understanding of hydrological and climatic variability (e.g. streamflow, precipitation and temperature) in space and time dimensions is essential to support management of water resources for humans and ecosystem needs. However, land use and climate changes caused by natural and/or human interactions may risk the life of societies by extreme shocks like floods and droughts. These effects may result in loss of lives and hinder the economic development of the society at large. Therefore, investigation of existence of trends in hydro-climatic variables provides important benefits for both operational and planning purposes of the present and future water resources at local and regional scales (see chapter 3).

Under these circumstances, good quality of hydro-climatic data is essential for the development of hydrological models to predict extreme flows, impacts of land use and climatic changes and for sustainable management of water resources in the Nile region (Kim and Kaluarachchi, 2008). In many parts of the globe, ground-based measurements of climate parameters and water fluxes are either missing or of poor quality for predicting and investigating changes in the hydrological system (Winsemius, 2009). This problem is common in developing countries, where logistic and financial constraints play a key role in data collection. Moreover, undervaluing the usefulness of such valuable information has made the data collection and data processing task more challenging. This could be one of the main factors, which challenges proper implementation of hydrological studies in developing countries.

Understanding of the catchment and its hydrologic processes is crucial for sustainable water resources management. Hydrological understanding enables improved conceptualisation of hydrological models, allows to quantify runoff components and to identify the dominant hydrological processes (Uhlenbrook et al., 2008). Therefore, stable environmental isotopes have been used to gain further insights about runoff generation mechanisms and identification of dominant hydrological processes in meso-scale catchments (e.g. Uhlenbrook et al., 2002, 2008; Tetzlaff et al., 2007b).

Environmental isotopes as tracers have a high potential for hydrological studies. Their application in hydrology started in the 1970s. Previous studies, e.g. Dincer et al. (1970), Sklash and Farvolden (1979) and Mosley (1979), have shown that the groundwater flow is often the dominant runoff component during the storm events in temperate climate. Since, then many advances have been seen in hydrological community by utilizing the environmental isotope methods in hydrology, in particular in the humid and temperate climate (Soulsby et al., 2000; Uhlenbrook et al., 2002; Hrachowitz et al., 2009). However, application of isotope techniques in the semi-arid and monsoonal climate of Africa is limited and largely unexplored (Mul et al., 2008; Wenninger et al., 2008; Hrachowitz et al., 2011a; Munyaneza et al., 2012). Numerous field experiments have been conducted in experimental research catchments worldwide for a better understanding of the hydrological processes using isotopes (see e.g. McDonnell et al., 1990; Uhlenbrook et al., 2002; McGlynn et al., 2004; Fenicia et al., 2008b; Birkel et al., 2010). Further studies explored the benefits of environmental isotopes as a tracer to estimate the mean residence time of water in a catchment (e.g. Rodgers et al., 2005a, b; Tetzlaff et al., 2009; Hrachowitz et al., 2011a, b). The potential of such tools have been explored for a better understanding of hydrological processes in the headwater of the Abay/Upper Blue Nile basin in this thesis (see Chapter, 6).

Hydrological processes are heterogeneous at all spatial and temporal scales (e.g. Blöschl and Sivapalan, 1995), and linking the processes in scaling relationships are a key to identify process controls on the appropriate spatio-temporal scales (Sivapalan, 2005; Didszun and Uhlenbrook, 2008). Many advances have been seen in hydrological process understanding particularly after the advent of the science initiative of predictions in the ungauged basins (PUB) that was launched by the International Association of Hydrological Science (IAHS) in 2003. The initiative aimed at devising and implementing a co-ordinated science programme to involve the scientific community, and to improve their capacity to achieve the premises towards new advances for making predictions in the ungauged basins (Sivapalan et al., 2003d). Consequently, a decade of predictions in the ungauged basins brought significant advancement in scientific understanding of hydrological processes, new approaches to data collection and model development, uncertainty analysis and model diagnostics, catchment classification and development of hydrological theories (Hrachowitz et al., 2013b).

Under the umbrella of the PUB initiative, numerous research work focused on process understanding to gain insights into the dominant hydrological processes using new additional data and new measurements, e.g. piezometric levels, soil moisture, tracer dynamics and geophysical methods (e.g. electrical resistivity tomography, ERT) (e.g. Uhlenbrook and Wenninger, 2006; Lehmann et al., 2007; Son and Sivapalan, 2007; Fenicia et al, 2008b; Winsemius et al. 2009). A number of process based studies emphasized that hydrological processes are influenced by threshold processes or behaviour (Blöschl and Zehe, 2005; Lehmann et al., 2007; Zehe and Sivapalan, 2009; Ali et al., 2013). For instance the "fill and spill mechanism" controlled by the bed rock topography is observed as a threshold behaviour of a sub-surface storm flow at hillslope scale (Spence and Woo, 2006; Tromp-van-Meerveld and McDonnell, 2006). Similar studies indicated that the spatial distribution of headwater storage as a threshold is crucial for determining which parts of a catchment contribute when to runoff generation (Spence et al. 2010; Phillips et al. 2011). A detailed overview of a concept of threshold behaviour in hydrological systems is given by Zehe and Sivapalan (2009).

Other studies elucidated that different landscape types entail distinct hydrological function and process dynamics, which was investigated at different spatio-temporal scales (e.g. McGlynn et al., 2004; Seibert et al., 2003; Buttle et al., 2005). Recently, Savenije (2010) suggested hydrologically meaningful landscape classification metrics that could be achieved by dividing the catchments in a semi-distributed manner that allows for the individual runoff processes to be assigned to different landscape units. Thus, allowing individual processes to be linked with distinct hydrological functioning.

Furthermore, Savenije (2010) suggested that for improved understanding of hydrological processes, catchments could be dissected into three main hydrological landscape units, i.e. plateaus, hillslopes and wetlands. The plateau is one landscape unit with modest slope, where groundwater is deep and evaporation excess deep percolation (DP) s the dominant runoff generation mechanism. The hillslope is the landscape unit where storage excess subsurface flow (SSF) is the dominant runoff mechanism, and a wetland is the landscape unit where saturation excess overland flow (SOF) is the dominant runoff mechanism. Renno et al. (2008) showed that the Height Above the Nearest Drainage (HAND) approach is a valuable tool to distinguish the hydrological landscape units. Gharari et al. (2011) demonstrated that HAND and slope derived from the digital elevation model appear to be the dominant topographical controls for such a process-based hydrological classification and that such a classification is useful for the development of process-based conceptual models.

## 1.3 Hydrological modelling

Hydrological models are useful tools for water resources assessment, understanding of hydrological processes and prediction of the impact of changes in land use and climate (Wagener et al., 2003). Over the past decades various rainfall-runoff models have been developed in different parts of the world. These models can be broadly classified based on the process description as physically based distributed models, conceptual models and data driven models (Beven, 2001). Few examples of physically based models include the SHE (System Hydrologique European) model (Abbott et al. 1986); CSIRO TOPOG Model (Vertesy et al. 1993); the IHDM model, (Claver and Wood, 1995); the HILLFLOW model, (Bronstert and Plate, 1997).

The models under a conceptual category include e.g. Sacramento Soil Moisture Accounting model (Burnash et al. 1973), TOPMODEL (Beven and Kirkby, 1979), TANK model (Sugawara, 1967), HBV model, Bergström, (1995); HYMOD model, Boyle, (2000) and Vrugt et al. (2003) among others. An overview for classification of hydrological model is described in more detail in, for instance, e.g. Singh (1995) or Beven (2001).

Klemes (1983) proposed two types of modelling approaches: a bottom-up and a top-down approach. Physically based distributed models are examples of a bottom-up modelling approach (Savenije, 2001; Sivapalan, 2003b). These models are largely based on the principles of physical processes based on continuity and the conservation of energy, mass and momentum. In this modelling approach, hydrological processes are modelled by introducing a large number of model parameters that supposed to be measurable at a plot or micro catchment scale, representing the different heterogeneities in the catchment. Since models built according to this approach are based on small scale theories of hydrologic response, they are very complex models, because they try to capture the details of process heterogeneity known at smaller spatial and temporal scales at the catchment scale (Sivapalan and Young, 2005). Consequently, these models often suffer from over-parameterisation and high predictive uncertainty problems (Uhlenbrook et al. 1999, Savenije, 2001 and 2009; Beven, 2002a, b; Sivapalan, 2005). The problem is that the hydrological processes at the smaller scale, which are often based on point measurements, might not explain the processes at the larger scale due to the heterogeneity in space (Blöschl and Sivapalan, 1995) and other process controls. The process scale representation using tracer based studies suggested that the dominant runoff generation processes are changing with scale (Didszun and Uhlenbrook, 2008).

In a top-down modelling approaches (e.g. conceptual models, data-driven models), the equations used to describe the physical processes often have (indirect) physical meaning but parameters are obtained through calibration. In this approach, the modelling procedure usually starts with a very simple model and progressively increased complexity through step-wise incorporation of process descriptions (Sivapalan, 2003b; Montanari et al., 2006). The FLEX models (flux exchange models) of Fenicia et al. (2006; 2008a, b) are good examples developed in a top-down fashion. In such modelling approach, testing and falsifying how a hydrological system works through testing hypotheses can be done by inclusion of additional information e.g., stable isotopes as tracers, groundwater level in addition to rainfall-runoff data (Seibert and McDonnell, 2002; Sieber and Uhlenbrook, 2005; Son and Sivapalan, 2007; Wissmeier and Uhlenbrook, 2007; Winsemius, 2009). Besides, the models developed are also based on empirical analysis with the inclusion of additional processes while achieving a more parsimonious model structure (e.g. Jothityangkoon et al., 2001; Atkinson et al., 2002; Zhang et al., 2008).

Recently, advances have been made in hydrological process understanding and modelling using topography driven, flexible, conceptual, semi-distributed model structures (Gharari et al., 2013; Gao et al., 2014). Hydrological processes represented in these models are based on the dominant processes related to the hydrological landscape units defined in section 2.1. This topography driven, a flexible model (FLEX-Topo approach) allows consistent comparison and testing of alternative model hypotheses (Gharari et al. 2011). Due to low parameterisation, accounting for the dominant runoff processes in different landscape units and its flexibility to test under different conditions the models have received recognition during the PUB research decade (Savenije, 2009, 2010; Hrachowitz et al., 2013b).

Data-driven models are based on extracting information that is implicitly contained in hydrological data. These models involve mathematical equations that do not rely on physically realistic principles such as mass, momentum or energy balance equations (Solomatine, 2011). The applications of such models depend on proper analysis of the input/output time series (Bowden et al., 2005). Artificial Neural Networks (ANN) are an example of data-driven model. The ANN models do not encapsulate hydrological knowledge that experts may have about the hydrological system (Corzo and Solomatine, 2007).

Simple conceptual water balance models (e.g. Conway, 2000) up to the complex process based distributed SWAT model (e.g. Setegne et al. 2010) have been applied in the Abay/Upper Blue Nile basin (see details in Chap. 4, Chap. 5 and Chap. 7). These models were applied on a larger scale and some of them were limited to the Lake Tana area and lack investigation of detail runoff generation processes. Furthermore, the models are very data demanding and contain a large number of model parameters, which increased the degree of freedom to reproduce the observations during calibration. In such conditions, flexible model structures with less input data, testing alternative hypotheses supported by complementary information like stable environmental isotopes offer a good opportunity for model improvement and gaining better process understanding (Wissmeier and Uhlenbrook, 2007; Fenicia et al., 2008).

## 1.4 Problem statement

The Abay/Upper Blue Nile basin constitutes the largest portion of the Nile water. Millions of people in the upstream and downstream of the river reach are reliant on the water availability. Consequently, a detailed hydrological study of the basin that identifies hydrological processes and runoff components is important to support water mangers and decision makers for sustainable water resources development in the region.

The basin is characterized by large temporal fluctuations in rainfall and runoff both at intra-annual and inter-annual scale as well as spatial variability in soil, geology, vegetation and topographic properties. Due to these heterogeneities, the hydrological processes in the basin are quite complex and highly variable in space and time. Human activities also influence the hydrological processes by interacting with the natural system. Intensive grazing, deforestation and improper and intense farming practices in unfavourable land are the major interferences in the basin, which alter the hydrological processes in the basin. These problems have been devastating in the Ethiopian highlands and cause erosion in agricultural land, reduction in crop yield, and they are responsible for the large siltation of irrigation canals and reservoirs in the downstream parts in Sudan and Egypt.

Unlike the lower part of the basin, the hydrology of the Abay/Upper Blue Nile basin is poorly understood. In general, limited hydrological and climatic data of sufficient quality have hindered sound research and hindered in-depth investigations of the basin hydrology. The spatial distribution of the existing observational network for hydro-climate data is inadequate, being too thinly spread and located in towns along the main roads only. It does not consider spatial variability in the mountains, in which precipitation amount is perceived to be high. In most cases the stream gauges are without proper rain gauge representation upstream of the catchment outlets. Thus, assessments of water resources through hydrological models or further hydrological studies are often difficult and not well enough supported by data. Under these circumstances, new observations of precipitation, water level, and stable isotopes including field process knowledge, in conjunction with the existing secondary data offers good opportunities for gaining futher insights in to the catchment functioning and better hydrological modelling in a data scarce area like the Abay/Upper Blue Nile basin.

## 1.5 Research objectives

The main objective of this research is to understand the hydrological processes and hydro-climatic variability in the Abay/Upper Blue Nile basin at various spatial and temporal scales. Therefore, the specific objectives are

1. To investigate the spatial and temporal long-term hydro-climatic trends in the basin;

2. To understand the water balance dynamics of the catchments within the basin at various spatial and temporal scales;

3. To evaluate the effects of past land use change on the hydrology of the selected agricultural meso-scale catchment Jedeb located in the headwater region of the Abay/Upper Blue Nile basin;

4. To characterise the spatial and temporal variability of stable environmental isotopes in two selected headwater catchments (Chemoga and Jedeb) in the Choke Mountains with the aim to identify mean residence times of water and the variable contribution of runoff components; and

5. To understand and model the rainfall-runoff processes in the two headwaters Chemoga and Jedeb using a process based modelling approach.

## 1.6 Dissertation structure

The thesis is organized in eight chapters. In the first chapter, scientific research issues pertinent to hydrological processes and modelling in tropical data scarce areas, the problem statement and the objectives of the thesis are presented.

In **Chapter 2**, the description of the study area, topography, climate, hydrology, land use, soil and geology is presented.

**Chapter 3** presents the assessment of long-term changes of the key hydro-climatic parameters (rainfall, temperature and river flows). The investigation is implemented in nine streamflow, thirteen precipitation and twelve temperature gauging stations. Statistical tests have been used to assess the significance of trends over different time periods.

**Chapter 4** presents the water balance study of twenty catchments using a top-down modelling approach based on Budyko's hypotheses. The method was applied on annual and monthly time scales to analyse the water balance in meso-scale catchments and at the larger scale of the Abay/Blue Nile at Ethiopian-Sudanese border.

In **Chapter 5**, modelling of the effects of land use change impacts on the Jedeb meso-scale headwater catchment of the Abay/Blue Nile basin is evaluated. Both, a statistical approach and a simple monthly conceptual model based on Budyko's hypothese presented in (Chap. 4) are used in the analysis.

**Chapter 6** presents the characterisation of stable environmental isotopes and its application in hydrology. The method is implemented in the Chemoga and Jedeb headwater catchments in the Abay/Upper Blue Nile basin.

**Chapter 7** presents the development of conceptual distributed model with the aim to understand the rainfall-runoff processes in the Chemoga and Jedeb headwater catchments. The purpose of this work is to gain insights about the hydrological processes in these fast responding headwater catchments. Different model representations with varying model complexity have been employed to test the suitability of the model structures. Parameters were conditioned using both discharge and environmental isotope information indicating the fraction of new and old water during the wet seasons.

Finally, **Chapter 8** summarises the findings of this thesis, concluding remarks, recommendation, and directions for further studies in the basin.

# CHAPTER 2

## Study area: the Abay/Upper Blue Nile river basin

The Abay/Upper Blue Nile River originates from Lake Tana in Ethiopia at an elevation of 1780 m a.s.l. The topography is generally rugged and mountainous and ranges from 489 m a.s.l. on the western part of the basin at Ethiopian–Sudan border to 4261 m a.s.l. on the Northern eastern part of the basin. It covers 173,686 km$^2$ catchment area upstream of the Ethiopian–Sudan border gauging station. (see fig. 2.1). Approximately 30 km downstream of Lake Tana, at the Tiss-Abay falls, the river falls into a deep gorge and travels about 940 km till the Ethiopian-Sudanese boarder (Conway, 1997).

### Climate

The climate in the Abay/Upper Blue Nile basin varies from humid to semi-arid to arid and it is mainly dominated by latitude and altitude. The influence of these factors determine a rich variety of local climates, ranging from hot and arid along the Ethiopia-Sudan border to temperate at the highlands and even humid-cold at the mountain peaks in Ethiopia. The mean annual temperature during the period 1961 to 1990 is 18.3°C with a seasonal variation of less than 2°C (Kim et al. 2008a). According to the data analysed in this thesis, the mean annual temperature ranges from 13°C in south eastern parts to 26°C in the lower areas of the south western part near to the Ethiopia-Sudan border for the period 1995-2004.

The Ethiopian National Meteorological Services Agency (NMSA) defines three seasons in Ethiopia: rainy season (June to September), dry season (October to January) and short rainy season (February to May) (NMSA, 1996). The short rains, originating from the Indian Ocean, are brought by south-east winds, while the heavy rains in the wet season originate mainly from the Atlantic Ocean and are related to south-west winds (BCEOM, 1999a; Seleshi and Zanke, 2004). The study by Camberlin (1997) reported that the monsoon activity in India is a major cause for summer rainfall variability in the East African highlands.

Some studies show that the climate in the basin is governed by the migration of the Inter Tropical Convergent Zone (ITCZ), which moves seasonally from the South to the North and back (e.g. Conway, 2000; Mohamed et al., 2005). The intra-annual rainfall variability has a mono-modal pattern.

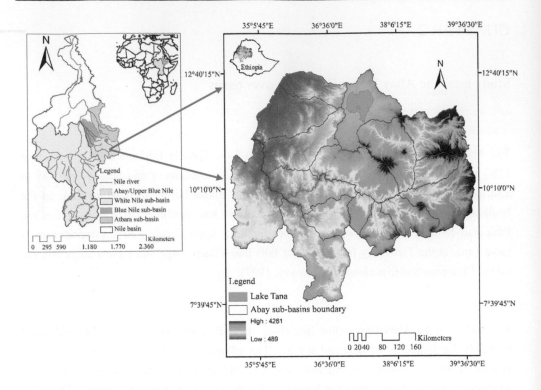

Figure 2.1: Location map of the Abay/Upper Blue Nile basin. The left figure shows the sub-basins of the Nile. The right figure shows the Abay/Upper Blue Nile sub-basin, which is the focus of this study.

Mean annual rainfall values estimated from eleven gauges range between 1148-1757 mm a$^{-1}$ during the period 1900-1998 and have a mean value of 1421 mm a$^{-1}$. 70% of the rainfall falls between June and September (Conway, 2000). Abtew et al. (2009) studied the spatial and temporal distribution of meteorological parameters in the basin. According to their study the mean annual rainfall is 1423 mm a$^{-1}$ for the period of 1960-1990. A recent study by Haile et al. (2009) showed that the variation of rainfall at the source of the Blue Nile River in Lake Tana sub-basin is affected by terrain elevation and distance to the centre of the Lake. Moreover, in their study it is indicated that the amount of nocturnal rainfall (rainfall during the night time) over the Lake shore was about 75% of the total rainfall and it is higher than the nocturnal rainfall over the mountainous areas. The average annual potential evaporation estimated using a multiple regression model based on latitude, longitude and the Thornthwaite formula to predict grid cell potential evaporation in the model amounts to 1100 mm a$^{-1}$, and it varies between 1200 and 1800 mm a$^{-1}$ in the upper and lower parts of the basin near to the Ethiopian-Sudan border, respectivily (Conway, 1997; Kim et al., 2008).

**Hydrology**

The Abay/Upper Blue Nile river in Ethiopia emerges from Lake Tana as an outflow and is fed by major tributaries like Weleka, Jemma, Beshilo, Muger, Guder, Fincha, Dedissa and Dabus joining the main stem of the river Abay at the left bank, and Bir, Beles, Chemoga and Jedeb are joining at the right bank. The Dinder and Rahad rivers originate from the Ethiopian highlands around the North-western part of Lake Tana sub-basin and are joining the Blue Nile in Sudan (see figure 2.1).

Lake Tana is the largest fresh water lake in Ethiopia having an area of 3100 km$^2$ and estimated storage volume of 28 km$^3$. The major tributaries feeding the lake are Gilgel Abay, Gumera, Rib and Megech. The outflow from the Lake contributes about 8% to the Abay/Blue Nile flow (Conway, 1997; Haile, 2010). The Lake has 73 km long and 68 km width. It has a mean depth of 9.53 m, while the deepest part is 14 m (ENTRO, 2007).

The flow in the Abay/Upper Blue Nile basin is highly seasonal pursuing the seasonality of rainfall in the basin. Most of the tributaries in the basin generate high runoff during high period of rainfall, June through September and decreases their flows or dry out in long dry season. The 5% and 95% flow exceedance extracted from daily flow duration curves and the mean monthly discharges at these key stations are shown in table 2.1 and figure 2.2, respectively.

Table 2.1: The 5% and 95% flow exceedance at the three key flow stations in the Abay/Upper Blue Nile basin.

| Flow station | Period of record | 5% exceedance ($m^3 s^{-1}$) | 95% exceedance ($m^3 s^{-1}$) |
|---|---|---|---|
| Abay at Bahir Dar | 1973-2006 | 399 | 9 |
| Abay at Kessie | 1960-2004 | 2347 | 11 |
| Blue Nile at El Diem | 1965-2010 | 5741 | 88 |

Based on the studies by Sutcliffe and Parks (1999), the mean annual flows of the Blue Nile is estimated to 48 km$^3$ a$^{-1}$ (1522 m$^3$ s$^{-1}$) for the period 1910-1995 at Khartoum gauging station. The flow analysis presented in this thesis indicate that the mean annual streamflow amounts to 47.48 km$^3$ a$^{-1}$ (1489 m$^3$ s$^{-1}$) discharge at the El-Diem gauging station in Sudan for the period 1965-2010, excluding the data gap and the whole 2005 data series, which is of suspicious quality.

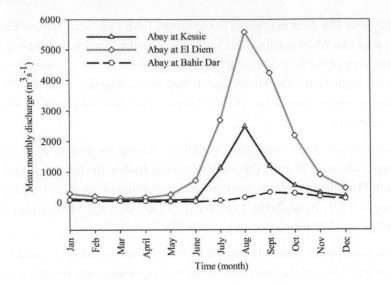

Figure 2.2: Mean monthly flows of Abay/Blue Nile at key gauging stations. At Bahir Dar for a period (1973-2006), at Kessie (1960-2004) and at El Diem (1965-2010).

The basin contributes 60% of the annual flow to the Nile River (e.g. UNESCO, 2004; Conway, 2005). Details of previous hydrological research work in the basin are provided in subsequent chapters of this thesis.

**Land use and soils**

ENTRO (2007) classified the land use/land cover based on the food and agriculture organization (FAO) classification system. The dominant land cover in the basin is rain-fed crops sedentary, i.e. a life style and farming system in which farmers raise crops permanently in the area in reverse to nomadic type of farming system, and grassland also equivalently dominant as rain-fed crops. This classification system demonstrated that rain-fed crops account for 25.8%, grasslands 25%, woodland 16.8%, shrubland 11.8% and cultivated land and semi-mechanized farms 10%. The remaining land cover accounts for less than 5%. The soil type in the basin is dominated by Vertisol and Nitisol types (53%). The Nitisols are deep non-swelling clay soils with favourable physical properties like drainage, workability and structure, while the Vertisols are characterised by swelling clay minerals with more unfavourable conditions (ENTRO, 2007).

**Geology**

The Abay/Upper Blue Nile river incision is caused by the Cenozoic basaltic uplifted land. Cenozoic basalts and ashes covered over two third of the basin. Crystalline basement rocks, volcanic rocks and sediments also make up the geology of the basin (Kebede, 2004). The basaltic cover is considered as perched groundwater systems with low storage and small aerial extent (BCEOM, 1998a). According to ENTRO (2007) the basin geology is characterised by basalt rocks, which are found in the Ethiopian highlands, while the lowlands are mainly composed of basement rocks and metamorphic rocks such as gneisses and marbles. Different literatures suggest that the geology is mainly dominated by volcanic rocks and Precambrian basement rocks with small areas of sedimentary rocks (Conway, 2000).

**Water resources development**

Until now no major, large-scale water resources developments projects have been implemented in the Abay/Upper Blue Nile basin. Conceivably, there are new plans on the implementation of water resources development projects underway. According to FDREMW (2002) potentially irrigable area in the basin is estimated to 760,000 ha. Nevertheless, the only major irrigation scheme in the basin is the Fincah sugar factory project that utilizes 6200 ha and 3100 ha of land is planned to be developed (BCEOM, 1999a). Besides, about 23,800 ha land is developed in small scale irrigation schemes in various parts of the basin (Arsano and Tamirat 2005). Recently, the multi-purpose hydro-power and irrigation projects in Beles and Neshi Rivers and the Anger-Dedissa irrigation schemes are the new irrigation developments blooming in the basin.

The largest hydro-power potential sites in Ethiopia are found in the Abay/Upper Blue Nile basin. The country's hydro-power potential is estimated to be 27,000-30,000 MW (Kloos and Legesse, 2010). Forty nine percent of the country's hydro power potential and more than half of the 75% dependable surface water available are in the Abay basin (Kloos and Legesse, 2010). Currently, the Ethiopian government has been launched a five year growth and transformation plan in order to boost the country's economy. Consequently, the Grand Ethiopian Renaissance Dam (GERD) development, which is located close to border to Sudan in the Abay/Upper Blue Nile basin, has started its construction in March 2011 (EEPCO, 2010). The project expected to develop 6000 MW to support the energy demand of the whole East African region. Furthermore, the cascade dam projects are also part of the future development plan upstream of the GERD. Recently, catchment management aiming at reducing soil erosion is one of the key development programs currently running in the basin.

# CHAPTER 3

## Hydro-climatic trends in the Abay/Upper Blue Nile basin[1]

Identification of trends in hydro-climatic variables has enormous use for planning and management of limited water resources and to set alternative strategies for future developments. Analyses of such trends have valuable use in particular to the shared water resources in the case of the Blue Nile river basin. This chapter presents the statistical methods for identifying the presence of trends in hydro-climatic variables; such as discharge, precipitation, and temperature data in the Abay/Upper Blue Nile basin. The results of the analyses show that trends and change point times varied considerably across stations and catchment to catchment with respect to the temperature and discharge series, respectively. However, precipitation did not show statistically significant trends both in annual and monthly time scales across all the investigated stations. Identified significant trends can help to make better planning decisions for water management. The details of the investigation have been described in subsequent sections of this chapter.

## 3.1 Introduction

In recent years significant progress has been made to study trends and variability of hydro-climate variables in different parts of the world. For example Hu et al. (2011) studied streamflow trends and climate linkages in the source region of the yellow river, China over the period 1959-2008. Their study concluded that the decrease in precipitation in wet season along with an increase in temperature causes the decrease in water availability for the downstream water user. Masih et al. (2010) studied streamflow trends and climate linkages in the Zagros Mountains, Iran over the period 1961-2001 and pointed out that most of the streamflow trends could be attribute to the change in precipitation. Love et al. (2010) showed that the rainfall and discharge from the northern part of Limpopo basin Zimbabwe depict declining trend. Abdul Aziz and Burn (2006) studied trends and variability in hydrological regime of Mackenzie River basin North Canada over different time period. Their study indicated that increasing flows for the winter season and the temperature data exhibit increasing trend in winter and spring seasons. Though, the precipitation data exhibit less well defined trend. Birsan et al. (2005) analyzed 48 catchments in Switzerland over the period

---

[1] *Based on*: Tekleab, S., Mohamed, Y, and Uhlenbrook, S. (2013). Hydro-climatic trends in the Abay/Upper Blue Nile basin, Ethiopia. Journal of Physics and Chemistry of the Earth. Doi: 10.1016/j.pce. 2013.04.017

1931-2000. They reported that strong relationship between streamflow trends and catchment characteristic suggests that mountain basin is most vulnerable environment from the point of view of climate change.

Hydrological and climatic variability are caused by multiple reasons. Studies show that anthropogenic climate change, modification in land use/clover, abstraction or change in water use are the main contributing factors for the alteration of hydrological and climatic variability (e.g. Pagano and Garen, 2004). Changes in climate in conjunction with the changes in physiographic characteristics of the catchment could influence the streamflow. Streamflow integrates and comprises spatial information about what is happening in the catchment and more appealing for detecting regional trends than point information like precipitation data (Birsan et al., 2005).

Assessment of long term variability of the transboundary Abay/Upper Blue Nile flow and patterns of hydro-climatic variables is crucial for sustainable water resources management and peace in the region (Kim et al., 2008). A number of hydrological studies have been conducted in the Abay/Upper Blue Nile to investigate basin water balance (e.g. Johnson and Curtis, 1994; Conway, 1997; Kebede et al., 2006; Mishra and Hata, 2006; Tekleab et al., 2011). A recent research addressed hydrological and runoff generation processes, e.g. Kim and Kaluarachchi (2008b), Rientjes et al. (2011a), Setegne et al. (2010), Uhlenbrook et al. (2010). Soil erosion and sedimentation studies were reported in Easton et al. (2010) and Betrie et al. (2011) among others. The literature showed many publications on land use/land cover change studies by Zeleke and Hurni, (2001); Hurni et al. (2005); Bewket and Sterk, (2005); Teferi et al. (2010), and Rientjes et al. (2011b). Few climate change impact studies were conducted by e.g. Kim et al. (2008a); Abdo et al. (2009); Elshamy et al. (2009) and Di Baldassarre et al. (2011).

Even with this good number of researches on various hydrological and environmental issues in the Abay/Upper Blue Nile basin, very little work has been done to investigate long-term trends of hydro-meteorological variables at catchment level. Conway and Hulme (1993) studied fluctuation in precipitation and runoff in Nile sub-basins and their impacts on Nile main discharges using historical data (1945-1984). They pointed out that higher inter-annual variability of runoff compared to precipitation, attributed to the fact that rainfall-runoff relationship is very sensitive to the fluctuation in precipitation. Furthermore, due to strong seasonality in rainfall; most of the tributaries in the basin drying out during the prolonged dry season (Conway, 2000). The rainy season covers only 3 to 5 month of the year. This is reflected in a very high seasonality of the river hydrograph. About 80% of the Blue Nile flow occurs in 4 month only. There are also increasing demands both nationally and at transboundary level for the same water resource.

Recently, Tesemma et al. (2010) studied trends of rainfall over the Abay/Upper Blue Nile basin, and streamflow at three gauging stations along the main stem of the river, which are at Bahir Dar (outlet of Lake Tana), Kessie and at the Ethiopian-Sudanese border (El-Diem), for the period 1964-2003. They reported that no statistically significant trends were observed in precipitation over the studied period. However, annual stream flow at Bahir Dar outflow from Lake Tana and Kessie showed significant increasing trends at 5% significance level, while no change was observed at El Diem station at Ethiopian–Sudanese border. During the main rainy season (June to September), significant increasing trends were observed at all three stations. The study by Eltahir (1996) indicates that the natural variability of annual flow of the Nile is associated with the El-Niño Southern oscillation (ENSO) phenomena. Similar studies also reported that the rainfall in the Ethiopian highlands during rainy season is tele-connected to the ENSO phenomena (Camberlin, 1995; Conway, 2000; Seleshi and Zanke, 2004).

In general, the investigations of trends of hydro-meteorological variables at sub-catchment level in previous studies were largely unexplored, though it is a very relevant issue for sustainable use of water resources at sub-catchment scale. The objective of this investigation is therefore to identify the presence of hydro-climatic trends in the Abay/Upper Blue Nile basin at sub-catchment level. This is a pre-requisite step to understand the cause and effect of trends and the links to water supply and use in the basin.

## 3.2 Study area and data sources

### 3.2.1 Study area

The location of the catchments and the climatic stations within the Abay/Upper Blue Nile basin are shown in figure 3.1. The catchment boundary and the drainage pattern have been delineated based on a 90*90 $m^2$ digital elevation model of the NASA Shuttle Radar Topographic Mission (SRTM), obtained from the Consortium for Spatial Information (CGIR_CSI) website (http://srtm.csi.cgiar.org).

### 3.2.2 Data sources

Stream flow data sets based on manual water level measurements (daily at 06:00 a.m. and 06:00 p.m.) for nine gauging stations in the Abay/Upper Blue Nile basin for different period (see Appendix A, Table A1) were collected from the Ethiopia Ministry of Water Resources and Energy. Similarly, monthly precipitation data for 13 stations and temperature data for 12 stations were obtained from the Ethiopian National Meteorological Agency.

In general, the hydro-climatic data is scanty in the basin and have many data gaps. Therefore, it is essential to devote careful screening and quality checks for all data before use in any hydrological analysis. Visual inspection and regression relations between neighbouring stations have been used to detect outliers and fill in missing gaps in the data series. For rainfall and temperature data the regression coefficient ($r^2$) ranges between 0.76 to 0.9. Large data gaps e.g. for the duration of one year and above, were excluded from the analysis. The selection of study catchments and gauging stations has been based on availability of data.

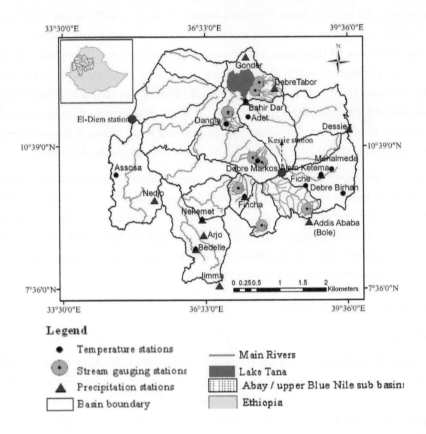

Figure 3.1: Location map of the study area and gauging stations. Numbers inside the green circles designates the river gauging stations. Names of the catchments referring to these numbers are provided in appendix A, table A1.

Missing values less than 5% for daily streamflow and 11% for monthly weather data have been used over the considered period (see Appendix-A, table A1 and A2).

## 3.3 Methodology

Investigating the presence of trends in hydro-meteorological variables is important for existing and future water resources developments in the Nile basin. To evaluate the existence of trends in precipitation, temperature, and stream flow time series of the Abay/Upper Blue Nile tributaries, the Mann-Kendall test hereinafter referred as MK test has been used (Mann,

1945; Kendall, 1975). Pettit tests have been employed to detect the change points in the time series. The periods of available data used in the analysis is given in appendix-A Table A1 and A2. Trends and change of points have been evaluated in different time period having varying length of available data series.

### 3.3.1 Mann-Kendall test

Mann-Kendall test is a rank based, non-parametric test, which has been widely used for detection of trends in time series (e.g. Zhang et al., 2001b; Love et al., 2010; Tesemma et al., 2010; Hu et al., 2011). It is applicable for non-normally distributed data, which are often encountered in hydrology or climatology, and is robust against outliers (Yue and Pilon, 2004; Hess et al., 2004). The null hypothesis is that a data series is serially independent and identically distributed with no trend.

The MK test statistic S is given by the formula:

$$S = \sum_{k=1}^{n-1} \sum_{j=k+1}^{n} \mathrm{sgn}(x_j - x_k) \tag{3.1}$$

Where $x_j$ and $x_k$ are the annual data values in years j and k, $j > k$ respectively.

$$\mathrm{sgn}(x_j - x_k) = \begin{cases} 1 & if\ x_j - x_k > 0 \\ 0 & if\ x_j - x_k = 0 \\ -1 & if\ x_j - x_k < 0 \end{cases} \tag{3.2}$$

The variance of (S) for independent and identically distributed random variables with no tied data and the standardized MK statistic Z follows the standard normal distribution with mean of zero and variance of one. The variance VAR(S) and Z are computed using equation (3.3) and (3.4), respectively. The trend results in this study have been evaluated at 5% significant level. This implies that the null hypothesis is rejected when $|Z| > 1.96$ in equation 3.4. Where, Z is the standard normal variate.

$$VAR(S) = \frac{n(n-1)\ (2n+5)}{18} \tag{3.3}$$

The presence of a statistically significant trend is evaluated using the Z value. A positive Z indicates an increasing trend, while a negative value indicates a decreasing trend. The MK test should be applied to serially independent or uncorrelated data (Helsel and Hirsch, 1992).

$$Z = \begin{cases} \dfrac{S-1}{\sqrt{VAR(S)}} & if\ S > 0 \\ 0 & if\ S = 0 \\ \dfrac{S+1}{\sqrt{VAR}} & if\ S < 0 \end{cases} \tag{3.4}$$

To correct the data for serial correlation, the procedure of trend free pre-whitening (TFPW), as described in Yue et al. (2002) and Yue et al. (2003), was used in this study. The steps used in TFPW start with de-trending the original series by removing linear trend in the data with slope

" i " estimated using equation (3.5). Then the de-trended series was computed using equation (3.6).

$$i = Median\left[\frac{x_j - x_k}{j - k}\right] \qquad (3.5)$$

For all k<j. (j =2,...., n and k=1,...., n-1)

$$Y_t = x_t - it \qquad (3.6)$$

Where $Y_t$ is the trended series, $x_t$ is the original data series value at time " t " and " i " is the slope.

The second step is accomplished by checking the presence of auto-correlation with lag-1 in the de-trended series using equation (3.7).

$$r_1 = \frac{\frac{1}{n-1}\sum_{t=1}^{n-1}(Y_t - \overline{Y_t})(Y_{t+1} - \overline{Y_t})}{\frac{1}{n}\sum_{t=1}^{n}(Y_t - \overline{Y_t})^2} \qquad (3.7)$$

Where $r_1$ is the lag-1 auto-correlation coefficient, n is number of years in the de-trended series, $Y_t$ is the trended series, and the over-bar indicates the mean of the trended series. The lag-1 auto-correlation coefficient at 5% significance level was evaluated to remove the auto-regressive part in the de-trended series. If the lag-1 auto-correlation coefficient is significant, then the auto-correlation part is removed from the series using equation (3.8) and if it is not significant, the MK test can be applied to the original $x_t$ data series.

$$Y_1 = Y_t - r_1 Y_{t-1} \qquad (3.8)$$

Where $Y_1$ is a series without auto-regressive part. Finally, the linear trend is added to the new series using equation (3.9) and MK test can be applied to this new series.

$$Y_2 = Y_1 + it \qquad (3.9)$$

Where: $Y_2$ is new series without auto-regressive and linear trend in the original data series.

### 3.3.2 Pettitt test for change point detection

To identify the date of a change point for hydro-climatic variables, the Pettitt test has been used (Pettitt, 1979). It is one of the important statistical methods to detect an abrupt change in a time series. It has been widely used to detect time change points in climatic and hydrological time series (e.g. Moraes et al., 1998; Mu et al., 2007; Love et al., 2010). It considers a sequence of random variables $x_1, x_2,..., x_T$, with a change point at $\tau$ ($x_t$ for t =1,2,..., $\tau$), and a common distribution function F1(x) and $x_t$ for t = $\tau$+1,..., T have a common distribution function F2(x), and F1(x) ≠ F2(x). The null hypothesis $H_0$: no change or $\tau$ = T is tested against the alternative hypothesis $H_a$: change or $1 \leq \tau < T$ using the non-parametric statistic $K_T = Max \left| U_{t,T} \right|, 1 \leq \tau \leq T$.

$$U_{t,T} = \sum_{i=1}^{t} \sum_{j=t+1}^{T} sgn(x_i - x_j) \tag{3.10}$$

$$sgn(x_i - x_j) = \begin{cases} 1 & if \ x_i - x_j > 0 \\ 0 & if \ x_i - x_j = 0 \\ -1 & if \ x_i - x_j < 0 \end{cases} \tag{3.11}$$

The significance level associated with $K_T$ is computed using equation (12).

$$P \cong 2 * exp\left[\frac{-6K_T^2}{T^3 + T^2}\right] \tag{3.12}$$

Where $P$ is the probability of detecting change of point. A '$P$' value less than significance level of 0.05 is identified as significant change of point in the time series considering 5% significance level. In this study a spread sheet model has been set up to detect the change of points in the data series.

## 3.4 Results and discussion

Summer is the main rainy season in the study area, which refer to the months June to September, spring is small rainy season March to May and winter is dry season October to February (NMSA, 1996). The serial correlation test indicates that none of the data series had statistically significant trends of serial correlation at 5% significance level. Hence, it has been used all the investigated data sets directly in the MK test.

### 3.4.1 Trends analysis of precipitation, temperature and streamflow

**Precipitation:**

Referring to the station provided in Appendix A, table A2, the results of the MK test for the 13 rainfall stations showed that, generally the trends are not significant both at monthly and seasonal scale for the majority of investigated climate stations. However, for some monthly values statistically significant decreasing trends were observed: in February in north central part of the basin at Debre Markos station with the MK statistic (Z = -2.1), in south east part of the basin at Jimma station (Z = -2.34) and Nedjo station (Z =-2.09), while rainfall in April showed significant increasing trend in North part Lake Tana sub basin at Debre Tabor (Z = 2.1) and statistically decreasing trend in August (Z = -2.0), while rainfall in June showed significant increasing trend in south east part of the basin at Fincha station ($Z = 2.88$) see the location of the stations in fig.3.1.

The results of trend analysis in precipitation found in this work are in agreement with past research work in the area (e.g. Conway and Hulme, 1993; Conway, 2000; Tesemma et al., 2010). They found that the monthly and seasonal areal rainfall did not show any evidences of distinctive statistically significant trends in the Abay/Upper Blue Nile basin. Seleshi and Zanke (2004) reported that no trends were observed in annual total, seasonal and rainy days over the central, northern and north-western Ethiopia in the period 1965-2002. Climate change studies also indicate that the GCM experiments show different direction of change in precipitation (wetter and drier) and they are not consistent but all agree on a temperature rises over the Abay/Upper Blue Nile basin (Elshamy et al., 2009; Nawaz et al., 2010).

**Temperature:**

Temperature data from 12 stations were investigated in the Abay/Upper Blue Nile basin over different time periods (see table A2 appendix A). The results of MK tests for the temperature stations showed that statistically significant increasing trends were observed in minimum, maximum and mean temperature for the majority of the stations in the basin. For the minimum temperature strong trends are prevalent in majority of the stations in all seasons (rainy, small rain and dry). However, spatially the stations found in south-west part of the basin showed negative trends in minimum, maximum and mean temperature (e.g. Nekemet and Assosa). Figure 3.2 presents mean annual and seasonal statistical trends of monthly mean of daily minimum, maximum and mean daily temperature for the investigated stations.

Strong increasing trends in temperature are apparent in the majority of the examined stations. Therefore, the increase in temperature particularly during the small rainy season has significant impact on the initial growing stage of the crop due to evaporation. The small rainy has significant economic importance for Ethiopia, since it triggers the growing season for most of the crops in the country (Seleshi and Camberlin, 2006). Dry spell analysis has been done at one station (Debre Markos) to see the trend of maximum number of dry spell length in the small rainy season. The dry spell analysis (maximum number of consecutive dry days with rainfall less than 1mm/day threshold) in this season in north central part of the basin at Debre Markos station show decreasing trend ($p = 0.66$) over the period 1954-2010, which is not statistically significant at 5% significance level. Thus, the wetting in spring season is potentially vulnerable to evaporation and may threaten the growth of the crops. A dry spell analysis for the whole Ethiopia over the period 1965-2002 done by Seleshi and Camberlin (2006) also indicated that the dry spell duration was not statistically significant for the Northern part of the country and statistically significant decreasing trends were seen in Eastern (Dire Dawa) and Rift valley (Awash), respectively.

Figure 3.3 illustrates the mean annual, rainy, small rainy and dry seasonal mean of average temperature trend magnitude for the selected temperature stations in Lake Tana sub basin at Bahir Dar and in the north central part of the basin at Debre Markos. The temperature has increased about 0.5, 0.3, 0.6, and 0.6°C/decade respectively for Bahir Dar station, and 0.29 0.2' 0.34 and 0.4°C/decade for Debre Markos station.

The results from present study are in agreement with previous climate change studies in the basin, who reported increasing trends of temperature (e.g. Abdo et al., 2009, Elshamy et al., 2009; Nawaz et al., 2010). Conway et al (2004) reported that over the period 1951 to 2002 the mean annual minimum and maximum temperature from Addis Ababa station showed increasing trend of 0.4°C/decade and 0.2°C/decade respectively.

**Streamflow:**

The results of MK test for the streamflow time series is summarized in Appendix A, Table A3. Table A3 presents the results of investigated stream flow trend for 9 Abay/Upper Blue Nile tributaries, which is characterized by different directions of trends. These heterogeneous results may suggest that the degree of human interventions and natural causes, which can be attributed to the change in stream flows are different across the examined catchments. For instance, figure 3.4 illustrates that the streamflow for the month January at Gilgel Abay in Lake Tana Sub basin and Jedeb catchment in north central part of the basin showed a statistically significant decreasing trends. In contrast Koga catchment adjacent to Gilgel Abay and Gumera catchments in Lake Tana sub basin depicted statistically increasing trends for the same month. Moreover, during the small rainy season statistically significant decreasing and increasing trends at Gilgel Abay and Guder catchments were identified. The mean annual stream flows for the examined catchment is characterized by a mixture of increasing and decreasing trends.

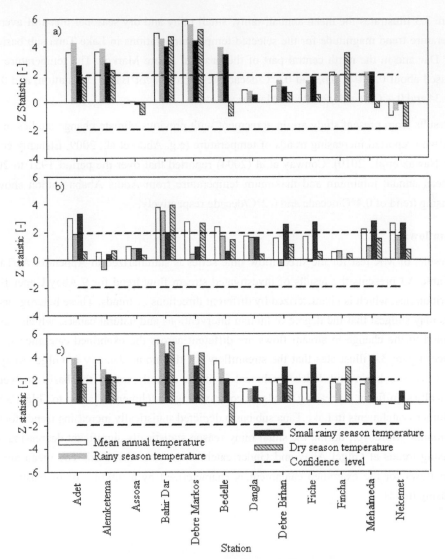

Figure 3.2: MK test Z statistics showing mean annual and seasonal temperature trends in the Abay/Upper Blue Nile climate stations: a) monthly mean of daily minimum temperature, b) monthly mean of daily maximum temperature, and c) monthly mean of daily mean temperature.

The MK test for mean annual stream flows showed statistically significant increasing trend in the south east part of the basin at the Neshi gauging station; the remaining 8 streams showed no significant trends at 5% confidence level.

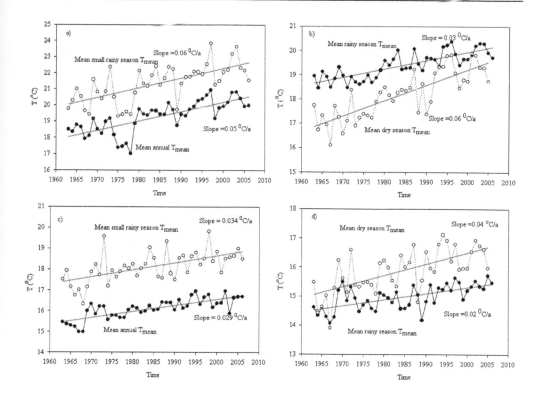

Figure 3.3: Mean annual and mean seasonal average temperature trend for selected stations in the Abay/Upper Blue Nile basin: a) and b) for Bahir Dar station, and c) and d) for Debre Markos station.

For monthly flows in the Gilgel Abay catchment, the results of the MK test indicate the presence of strong decreasing trends for the majority of the hydrological variables except the flow in June, which depict significant increasing trend. As expected the significant decreasing trend during dry season flow results in significant decreasing trend in 1 and 7-day annual minimum flows. Comparing the two adjacent catchments Gilgel Abay and Koga, demonstrate that the trend test results indicate that statistically significant decreasing trends in Gilgel Abay for most of the dry months (Nov, Dec, Jan, Feb) could not be detected in the Koga catchment. The most likely reason for this difference could be the existence of marshland and dambos in the Koga catchment, which create a smoothing effect in runoff distribution and hinder to detect the changes at the gauging station.

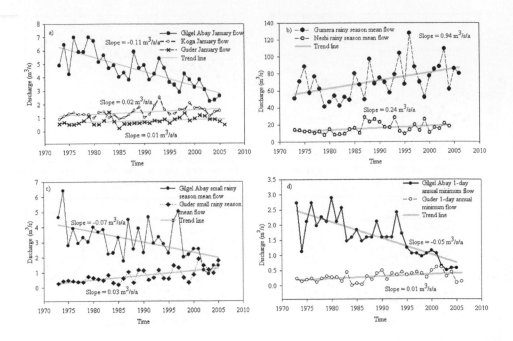

Figure 3.4: Time series plots for a) January flow in Gilgel Abay, Koga and Guder catchment, b) Rainy season mean flow in Gumera and Neshi catchment, c) Small rainy season mean flow in Gilgel Abay and Guder catchment, and d) 1-day annual minimum flow in Gilgel Abay and Guder catchment.

Moreover, large water storage and the presence of marshlands in the Koga catchment made the base flow production larger (Uhlenbrook et al., 2010). Study by Gebrehiwot et al. (2010) indicates that despite the reduction in the forest area, between 1960 and 2002 in the Koga catchment, the hydrological record did not reveal changes in low flow regime. This is due to the fact that the effects of deforestation have been buffered by the wetlands in the catchment. Hence, the streamflows in most of the months during dry season showed increasing trends.

For the Rib catchment in the Lake Tana sub basin and Chemoga catchment in the north central part of the basin, no discernible significant trends were observed for all hydrological variables at 5% level. However, monthly flow in February and 1-day annual minimum flow display significant decreasing trend at 10% level. It has to be noted that the majority of the trends showed decreasing trends. Similarly for the Koga and Muger catchment in the south east part of the basin, only the flow in Jan and 1-day minimum flow indicated significant increasing and decreasing trends respectively. Besides, increasing trend direction in Koga and decreasing trends in Muger catchments are dominating in most of the hydrological variables.

For the Jedeb catchment, the significant decreasing trends for monthly flows during dry season (Dec, Jan, and Feb) could be related to water abstraction for irrigation. Approximately 2.5 km upstream of the gauging station, a diversion weir has been constructed, in1996, but the abstractions have not been recorded. As can be seen in figure 3.5, the Jan flow exhibit significant decreasing trends and it is more pronounced particularly after 1996 onwards. Similarly, for the months December and February significant decreasing trends were observed.

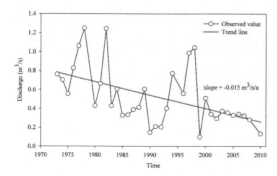

Figure 3.5: Time series plot of January flow for Jedeb catchment.

The trend results of the Blue Nile at the El-Diem gauging station at the Ethiopia Sudan border at the larger spatial scale (approximately 176,000 km$^2$) over the period 1965-2010, interestingly all the hydrological variables indicate increasing trends (results are not shown). Particularly, the monthly flows from March to July, small rainy season mean flow, and 1 and 7-day annual minimum flows exhibit statistically significant increasing trends at 5% level. However, it was not possible to make a comparison between the trend results of the tributaries with the El-Diem trend results due to the fact that many of the tributaries were not investigated for trends due to limited data availability in the basin.

Overall, the comparison of different catchments for different time period leads to observe different trend results. Thus, trend detection test results are quite depend on the number of years of data record length and hence, one can get different results for different time periods. Moreover, different results for the investigated catchments might be related to difference in contributory factors both human interferences and physiographic characteristics like climate, vegetation, soil, geologic and topography lead to differences in responsive characteristics of the catchments.

### 3.4.2 Analysis of change points

The Pettitt test identified different change points, with statistically significant change in means of the time series as also confirmed by t test (results not presented here). Here it has been presented only the change points for temperature, and discharge on mean annual and seasonal time scale. For the rainfall stations change point detection results were insignificant and are not presented here.

### Detection of change point for temperature time series

The results from Pettitt test reveal that, for the majority of the stations statistically significant increasing shifts have been detected (Appendix-A, table A4). From the table A4 it can be observed that changes of points are different for different stations. Due to different data record length for detection of change point, we could not compare the results directly. Even for the same data set (1963-2006) for Bahir Dar and Debre Markos stations, the time for change points is quite different. Though, for the dry season the mean temperature change point was detected in 1983 at these both stations. It is apparent that the change point for those stations occurred around early 1980s and 1990s and mid 1980s and 1990s.

### Detection of change of point in streamflow time series

The change point detection using Pettit test identified several change point for the nine Abay/Upper Blue Nile catchments (see Appendix-A, table A5). For the selected catchments the change point detection is illustrated in figure 3.6.

The results from the Pettit test showed mean annual stream flows in Koga, Gumera and Neshi catchments exhibited an upward shift, occurred around 1995, 1987 and 1986, respectively. While the mean annual flow at Jedeb showed downward shift that occurred around 1983. The remaining catchments, showed an upward shift, which is dominating and downward shifts, though they are not statistically significant at 5% level.

For the different months, the results of change point analysis reveal that both upward and downward shifts have been detected. In general, an abrupt change in upward direction was prevalent in the months of the dry season and the small rainy season, and the change points in the downward direction were detected in four out of nine gauging stations during the rainy season.

For 1-day minimum flow, statistically significant upward shift was found at stations Gumera and Guder around the years, 1990 and 1998, respectively. Conversely, statistically significant downward shift for 1-Day annual minimum flow was detected in Gilgel Abay station around the year 1994. For 7-day annual minimum flow, statistically significant upward shift was detected at Gumera and Guder catchment around the year 1990 and 1998, respectively and a significant downward shift were found at Gilgel Abay catchment around the year 1994.

The extreme 1-day annual maximum flows exhibit statistically significant upward shifts at the Jedeb, Guder and Neshi catchments around the year 2002, 1988 and 1986, respectively. Conversely, for 1-day and 7-day annual maximum flows, statistically downward shifts were detected around the years 1980 and 1981 in the Rib. In the Chemoga catchment, for the 7-day annual maximum flows a downward shift was detected around the year 1994. For 7-day annual maximum flows, a statistically upward shift was detected in the Neshi catchment around the year 1986.

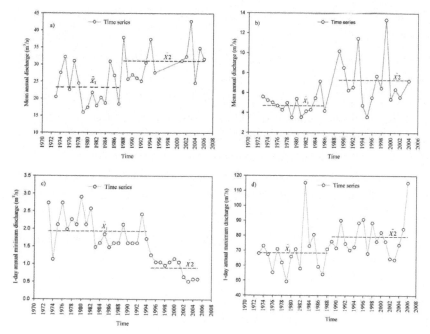

Figure 3.6: Time series plots for the change of time detection a) Gumera catchment mean annual discharge, b) Neshi catchment mean annual discharge, c) Gilgel Abay catchment 1-day annual minimum discharge, and d) Guder catchment 1-day annual maximum discharge. The dash lines represented by $\bar{x}_1$ and $\bar{x}_2$ are the mean of the time series before and after the change point respectively.

## 3.5 Conclusions

The trend and change point analysis were carried out for 13 precipitation, 12 temperature, and 9 stream gauging stations with in the Abay/Upper Blue Nile basin over different time periods. The selection of stations and records was determined by data availability and quality. The results in general are characterized by statistically significant increasing trends in temperature, no significant changes in mean annual or mean seasonal precipitation and both statistically significant increasing and decreasing streamflow trends in seasonal and extreme flow.

From the climate variables the minimum temperature showed more significant increase than the maximum and mean temperatures. The increases are more pronounced in small rainy and dry seasons than during the rainy season. Over different periods of investigation, the precipitation in February at 12 stations showed decreasing direction and statistically significant in the north central part of the basin at Debre Markos, and south east part of the basin at Jimma and Nedjo. In contrast precipitation in April and June in Lake Tana sub basin at Debre Tabor and south east part of the basin in June at Fincha exhibit statistically significant increasing trend.

For the hydrological variables in the Koga, Rib, Jedeb and Chemoga catchments no significant changes are seen on mean annual, seasonal and extreme flows. In contrary in the Gumera and Neshi catchments, the rainy season mean flows exhibit significant increasing trends. However, the small rainy season mean flow show significant decreasing trend in Gilgel Abay and significant increasing trends in Gumera and Guder catchment. 1- and 7-day annual minimum flows show significant decreasing trend in Gilgel Abay and Muger catchment. In contrast 1-day annual minimum flow and 1-day annual maximum flow in Guder show significant increasing trend. Besides, the1-and 7-day annual maximum flow display significant increasing trends in the Neshi catchment.

The Pettit test results for the change of point also display different times which are not consistent for both temperature and streamflows data series. For the majority of the streamflow gauging stations the 1990s are the dominating change time. In general there is a tendency of increasing trend and change point sign for the small rainy and dry seasons and 1- and 7-day annual minimum flow in the investigated catchments.

It is speculated that the changes for the examined hydro-meteorological variables might be related to both changes in climate variability either local or global and land use/land cover change over the investigated periods. Nevertheless, quantification of the causes attributed to the changes in stream flows either from anthropogenic activities (land use/land cover change) or climate change or climate variability are beyond the focus of this study. Therefore, there is a need for more research linking land use change detection using remote sensing and statistical analysis of hydro-meteorological variables. Moreover, it is recommended that quantifying the effects of land use/land cover and climate change using rainfall-runoff modelling is crucial for proper utilisation of the limited water resources in the region.

Moreover, it is stressed that limited spatial coverage of weather stations and streamflow measurements on the one hand, and the poor qualities of some data sets on the other hand, are the major limitations for a better understanding of environmental changes in the Abay/Upper Blue Nile basin. Attention should be paid in hydro-climatic data collection and processing to enable to study the impacts of climate variability, climate and land cover changes with certainty on existing, ongoing and future water resources development projects in the basin. This is a perquisite for peace and sustainable development in the region.

# CHAPTER 4

Water balance modelling of the Upper Blue Nile catchments using a top-down approach[1]

This chapter presents the water balance analysis of the Upper Blue Nile catchments based on Budyko hypotheses using a top-down modelling approach. The study covers various spatial and temporal scales. First, the annual water balance of the twenty catchments were analysed based on the available water and atmospheric demand. Simple Budyko curve represented by the aridity index (the ratio of annual potential evaporation to annual precipitation), could not well captured the observed annual runoff and annual evaporation. The model then improved by take in to account the soil moisture dynamics in the root zone at monthly time scale. The multi-objective model calibration strategy and the Generalized Likelihood Uncertainty Estimation (GLUE) were employed to condition the model parameters and uncertainty assessment. The model was calibrated and validated against the observed runoff data and it shows good performance for the twenty study catchments. The details of the method and the results are discussed in more detail in the following sections.

## 4.1 Introduction

Despite its 60% of annual flow contribution to the Nile river (e.g. UNESCO, 2004; Conway, 2005), the research in the Blue basin has suffered from limited hydrological and climatic data availability, which hampers an in-depth study of the hydrology of the basin.

The hydrology of the Upper Blue Nile basin was studied using a simple water balance models described earlier in Chapter 3. Most of these studies in the past were conducted on large scale to analyse the flow at the outlet at the Ethiopia-Sudan border. An understanding of the processes at sub-catchment level is generally lacking. Moreover, as a part of model uncertainty and parameter identifiability studies, multi-objective calibration searching for optimal parameter sets towards different objective functions was also missing in previous

---

[1] *Based on*: Tekleab, S., Uhlenbrook, S., Mohamed, Y., Savenije, H.H.G., Temsegen, M, and Wenninger, J. (2011). Water balance modelling of the upper Blue Nile catchments using a top-down approach. Hydrology and Earth System Science 15: 2179–2193, doi: 10.5194/hess-15-2179.

studies. Another limitation is that most of the hydrological studies in the Blue Nile have been conducted in the Lake Tana sub-basin. For example, the applications of the SWAT model in Lake Tana sub-basins (Setegne et al., 2010). Kebede et al. (2006) studied the water balance of Lake Tana and its sensitivity to fluctuations in rainfall. Rientjes et al. (2011a) applied HBV model in the Lake Tana sub basin to study ungauged catchment contributions to the Lake Water balance.

Uhlenbrook et al. (2010) studied the hydrological dynamics and processes of Gilgel Abay and Koga catchments using also the HBV model. Assessments of climate change impacts on hydrology of the Gilgel Abay catchment also indicate that the catchment is sensitive to climate change especially to changes in rainfall (Abdo et al., 2009). Gebrehiwot et al. (2010) studied the relation of forest cover and streamflow in the headwater Koga catchment through satellite imagery and community perception. They reported that the effect of deforestation for the past four decades did not show any significant change in the flow regime.

Assessment of catchment water balance is a pre-requisite to understand the key processes of the hydrologic cycle. However, the challenge is more distinct in developing countries, where data on climate and runoff is scarce as in the case of Upper Blue Nile basin. In such cases, a water balance study can provide insights into the hydrological behaviour of a catchment and can be used to identify changes in main hydrological processes (Zhang et al., 1999). In order to analyze the catchment water balance Budyko (1974) developed a framework linking climate to evaporation and runoff from a catchment. He developed an empirical relationship between the ratio of mean annual actual evaporation to mean annual rainfall and mean annual dryness index of the catchment.

The Budyko hypothesis has been widely applied in the catchments of the former Union of Soviet Socialist Republics (USSR). Similar studies were conducted worldwide using Budyko's framework (Sankarasubramania and Vogel, 2003; Zhang et al., 2004; Donohue et al., 2007; Gerrits et al., 2009; Potter and Zhang, 2009; Yang et al., 2009). All these studies improved Budyko framework by including additional processes. Zhang et al. (2001a; 2008) and Yang et al. (2007) suggested that by assuming negligible storage effects for long term mean (> 5 year) the annual aridity index $\left(\dfrac{E_P}{P}\right)$ controls partitioning of precipitation (P) to evaporation (E) and runoff (Q). $(E_P)$ is the potential evaporation, while E is actual evaporation. By evaporation we mean all forms of water changes from liquid to vapour, i.e., soil and open water evaporation plus transpiration and interception evaporation.

This is often termed evapotranspiration or total evaporation in the literature. However, Sankarasubramania and Vogel (2003) argued that the aridity index is not the only variable controlling the water balance at annual time scale and that the evaporation ratio (E/P) is related to soil moisture storage as well. Their results improved by including a soil moisture storage index, which could be derived from the 'abcd' watershed model. The abcd watershed model is a nonlinear water balance model which uses precipitation and potential evaporation as input, producing streamflow as output.

The model has four parameters a, b, c and d. The parameter 'a' represents the tendency of runoff to occur before the soil is fully saturated. The parameter 'b' is an upper limit on the sum of evaporation and soil moisture storage. The parameter 'c' represents the fraction of streamflow which, arises from groundwater. The parameter 'd' represents the base flow recession constant (Sankarasubramania and Vogel 2003). Besides, they classified catchments in the US based on the aridity index range of 0-0.33 as humid, 0.33-1 as semi-humid, 1-2 as temperate, 2-3 as semi-arid and 3-7 as arid. Milly (1994) showed that the spatial distribution of soil moisture holding capacity and temporal pattern of rainfall can affect catchment evaporation, but could be of small influence on annual time scale.

Obviously, spatial and temporal variability of vegetation affects evaporation and hence the water balance. Thus, it is more important to include the soil moisture storage for smaller spatial and temporal scales (Donohue et al., 2007). Zhang et al. (2004) hypothesized that the plant available water capacity coefficient reflects the effect of vegetation on the water balance. They developed a two parameter model which relates the mean annual evaporation to rainfall, potential evaporation and plant available water capacity to quantify the effect of long term vegetation change on mean annual evaporation ($\bar{E}$) in 250 catchments worldwide and found encouraging results. Inspired by the work of Fu (1981), Yang et al. (2007) analyzed the spatio-temporal variability of annual evaporation and runoff for 108 arid/semi-arid catchments in China and explored both regional and inter-annual variability in annual water balance and confirmed that the Fu (1981) equation can provide a full picture of the evaporation mechanism at the annual timescale.

The distinct feature of the investigation as compared to the previous studies is that, we learned from the data starting with simple annual model in different sub-catchments with in the Upper Blue Nile basin based on Budyko framework and model complexity is increased to monthly time scale. The monthly water balance model developed by Zhang et al. (2008), which is based on the Budyko hypotheses, has been tested in 250 catchments in Australia with different rainfall regimes across various geographical region and they obtained encouraging results. This model has applied in the Upper Blue Nile catchments due to its parsimony, only having four physically meaningful parameters and its versatility to predict streamflow and to investigate impacts of vegetation cover change on stream flow. Fang et al. (2009) applied the model in Australian and South African catchments on a monthly time scale and meso-scale and to large scale to study land cover change impacts on streamflow.

Moreover, in data scares environment like the Upper Blue Nile basin, complex models which require more input data and large number of parameters are not recommended. The model used in this study has only four parameters. The most commonly used hydrological models which have been applied in the Blue Nile basin are conceptual semi-distributed model like the Soil and Water Assessment Tool (SWAT) (more than 20 parameters, many temporal variable) and the conceptual HBV model (Hydrologiska Byrans Vattenbalansavdelning) has (12 parameters). They have many more parameters than the four parameters model used in this study. Therefore, in terms of over parameterization, which is the major cause of equifinality, the previous models were more prone to equifinality problem than the model we used in this study.

The objective of this study is building on the work of Budyko (1974), Fu (1981), Zhang et al. (2004, 2008) to investigate the water balance dynamics of twenty catchments in the Upper Blue Nile on temporal scales of monthly and annual and spatial scales of meso-scale (catchment area between 10-1000 km$^2$) to large scale (catchment area > 1000 km$^2$). Consequently, the basis for predicting water balance parameters in ungauged basins is laid.

## 4.2 Study area and input data

### 4.2.1 Study area

The location of the studied catchments within the Upper Blue Nile basin is shown in figure 4.1. The details about the climate of the basin are described in chapter 2.

### 4.2.2 Input data

Monthly stream flow time series of 20 rivers covering the period 1995–2004 have been collected from the Ministry of Water Resources Ethiopia, Department of Hydrology. The quality of the input data has been checked based on comparison graphs of neighbouring stations and also double mass analyses were carried out to check the consistency of the time series on a monthly basis. The missing data were filled in using regression analysis. Monthly meteorological data for the same period were obtained from the Ethiopian National Meteorological Service Agency (ENMSA). The data comprises precipitation from 48 stations and temperature from 38 stations. Potential evaporation was computed using Hargreaves method with minimum and maximum average monthly temperature as input data (Hargreaves and Samani, 1982). This method was selected due to the fact that meteorological data in the region is scarce, which limits the possibility to use different methods for the computation of potential evaporation. However, the Penman-Monteith method, which has been applied successfully in different parts of the world, was compared with other methods and is accepted as the preferred method for computing potential evaporation from meteorological data (Allen, et al., 1998; Zhao et al., 2005). The Hargreaves model was recommended for the computation of potential evaporation, if only the maximum and minimum air temperatures are available (Allen et al., 1998). Hargreaves and Allen (2003) also reported that the results computing the monthly potential evaporation estimates obtained using Hargreaves method were satisfactory. For example Sankarasubramania and Vogel (2003) used Hargreaves model to estimate the monthly potential evaporation in 1337 catchments in U.S. Table 4.1 presents basic hydro-meteorological characteristics of the twenty study catchments.

Table 4.1: Hydro-meteorological characteristics of the investigated twenty upper Blue Nile catchments (1995-2004); P, $E_P$, Q and E stand for basin precipitation, potential evaporation, discharge and actual evaporation, respectively.

|  |  |  | Long term mean annual values (mm a$^{-1}$) | | | |
| --- | --- | --- | --- | --- | --- | --- |
| Catchment number | Catchment name | Area (km$^2$) | P | $E_P$ | Q | E |
| 1 | Megech | 511 | 1138 | 1683 | 421 | 717 |
| 2 | Rib | 1289 | 1288 | 1583 | 349 | 939 |
| 3 | Gumera | 1269 | 1330 | 1671 | 841 | 488 |
| 4 | Beles | 3114 | 1402 | 1826 | 643 | 759 |
| 5 | Koga | 295 | 1286 | 1751 | 606 | 679 |
| 6 | Gilgel Abay | 1659 | 1724 | 1719 | 1043 | 681 |
| 7 | Gilgel Beles | 483 | 1703 | 1741 | 897 | 806 |
| 8 | Dura | 592 | 2061 | 1788 | 1004 | 1056 |
| 9 | Fetam | 200 | 2357 | 1619 | 1400 | 956 |
| 10 | Birr | 978 | 1644 | 1677 | 553 | 1091 |
| 11 | Temcha | 425 | 1409 | 1512 | 623 | 785 |
| 12 | Jedeb | 296 | 1449 | 1454 | 861 | 587 |
| 13 | Chemoga | 358 | 1441 | 1454 | 461 | 980 |
| 14 | Robigumero | 938 | 1052 | 1362 | 289 | 762 |
| 15 | Robijida | 743 | 1002 | 1379 | 222 | 780 |
| 16 | Muger | 486 | 1215 | 1654 | 480 | 735 |
| 17 | Guder | 512 | 1734 | 1744 | 713 | 1020 |
| 18 | Neshi | 327 | 1654 | 1416 | 686 | 967 |
| 19 | Uke | 247 | 1957 | 1488 | 1355 | 602 |
| 20 | Didessa | 9672 | 1538 | 1639 | 334 | 1204 |

monthly potential evaporation estimates obtained using Hargreaves method, water balance, For example, Smith (Smith 1993) and Allen et al. (2005) used Hargreaves method to estimate the reference potential evaporation in to FAO reference by ...

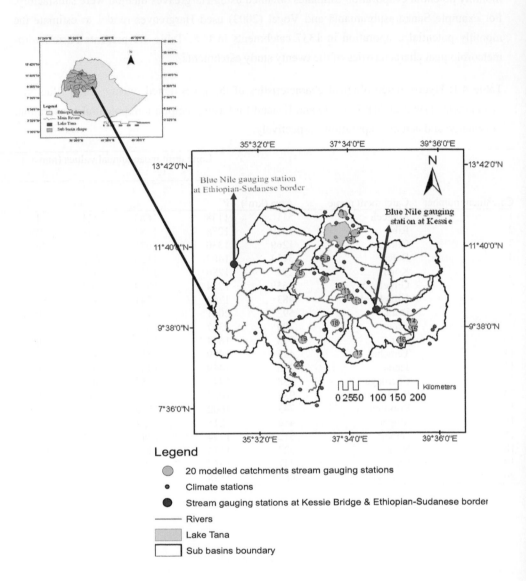

Figure 4.1: Study area; numbers indicate the location of gauging stations of modelled catchments.

## 4.3 Methodology

### 4.3.1 The Budyko framework

Budyko-type curves were developed by many researchers in the past (Schrciber, 1904; Ol'dekop, 1911; Turc, 1954; Pike, 1964; Fu, 1981). The assumptions inherent to the Budyko framework are:

1. Considering a long period of time ($\tau \geq 5$ years), the storage variation in catchments may be disregarded (i.e. $\Delta S \approx 0$).

2. Long term annual evaporation from a catchment is determined by rainfall and atmospheric demand. As a result, under very dry conditions, potential evaporation may exceed precipitation, and actual evaporation approaches precipitation. These relations can be written as $\dfrac{Q}{P} \to 0$, $\dfrac{E}{P} \to 1$, and $\dfrac{E_P}{P} = \to \infty$ similarly, under very wet conditions, precipitation exceeds potential evaporation, and actual evaporation asymptotically approaches the potential evaporation $E \to E_P$, and $\dfrac{E_P}{P} = \to 0$.

The Budyko (1974) equation can be written as

$$\frac{E}{P} = \left[ \phi \tanh(\frac{1}{\phi})(1 - \exp^{-\phi}) \right]^{0.5}$$
(4.1)

$\phi$ is the aridity index, which is the ratio of mean annual potential evaporation to mean annual rainfall. The Budyko curve representing the energy and water limit asymptotes are shown in figure 4.2.

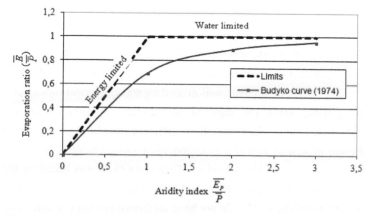

Figure 4.2: Budyko curve representing evaporation ratio is a function of aridity index.

### 4.3.2 Catchment water balance model at annual time scale

To study catchment water balance hydro-meteorological variables have been prepared for the model as an input. The average precipitation and potential evaporation for each twenty catchments have been estimated using Thiessen polygon method. Figure 4.1 illustrates the hydro-meteorological location used in this study. Due to scares data availability with many gaps in the time series, only twenty catchments with relatively having good quality of streamflow, precipitation and temperature measurements were used in this study. The water balance equation of a catchment at annual time scale can be written as:

$$\frac{dS}{dt} = P - Q - E \tag{4.2}$$

Where: $P$ is precipitation (mm a$^{-1}$)

$Q$ is the total runoff (mm a$^{-1}$), i.e. sum of surface runoff, interflow and baseflow

$E$ is actual evaporation (mm a$^{-1}$)

$\frac{dS}{dt}$ is storage change per time step (mm a$^{-1}$)

By assuming storage fluctuations are negligible over long time scales, i.e. $\tau \geq 5$ years, equation (4.2) reduces to:

$$P - Q - E = 0 \tag{4.3}$$

Equation (4.3) is known as a steady state or equilibrium water balance, which is controlled by available water and atmospheric demand controlled by available energy (Budyko, 1974; Fu, 1981; Zhang et al., 2004; Zhang et al., 2008).

Among the different of Budyko-type curves, we used equation (4.4) given by:

$$\frac{E}{P} = 1 + \frac{E_P}{P} - \left[1 + \left(\frac{E_P}{P}\right)^w\right]^{1/w} \tag{4.4}$$

Details of the derivation of this equation can be found in (Zhang et al., 2004). This equation has one annual parameter $w$[-], which is a coefficient representing the integrated effects of catchment characteristics such as vegetation cover, soil properties and catchment topography on the water balance (Zhang et al., 2004). This equation would enable to model individual catchments on annual basis. Even though, the results from different Budyko-type curves are not presented here, prediction of annual runoff and evaporation using equation (4.4) was better than the other Budyko-type curves. The selection of equation (4.4) was based on the Nash and Sutcliffe efficiency criterion/measure (Nash and Sutcliffe, 1970). Moreover, most of Budyko-type curves including Budyko (1974) do not have calibrated parameters and could not be applied on individual catchments (Potter and Zhang, 2009).

The two performance measures Nash and Sutcliffe, (1970) coefficient of efficiency ($E_{NS}$) and root mean squared error (RMSE) were used in the annual model, and are defined as follows:

$$E_{NS} = (1) - \frac{\sum\limits_{i=1}^{n}(Q_{sim,i} - Q_{obs,i})^2}{\sum\limits_{i=1}^{n}(Q_{obs,i} - \overline{Q}_{obs,i})^2}$$

(4.5)

$$RMSE = \sqrt{1/n * \sum\limits_{i=1}^{n}(Q_{obs,i} - Q_{sim,i})^2}$$

(4.6)

Where $Q_{sim,\ i}$ is the simulated streamflow at time $i$ [mma$^{-1}$], $Q_{obs,i}$ is the observed streamflow at time $i$ [mma$^{-1}$], $n$ is the number of time steps in the calibration period and the over-bar indicates the mean of observed streamflow.

### 4.3.3 Catchment water balance model at monthly time scale

The dynamic water balance model developed by Zhang et al. (2008) has been used to simulate the monthly streamflow. The model has four parameters describing direct runoff behaviour $\alpha_1$ [-], evaporation efficiency $\alpha_2$ [-], catchment storage capacity $S_{max}$ [mm], and slow flow component $d$ [1/month]. Zhang et al. (2008) stated that each parameter can be interpreted physically. For example the parameter $\alpha_1$ represents catchment rainfall retention efficiency and an increase in parameter $\alpha_1$ implies higher rainfall retention and less direct runoff. The maximum soil water storage in the root zone ($S_{max}$) relates to soil and vegetation characteristics of the catchment. The parameter $\alpha_2$ relates to the evaporation efficiency, a higher value implies a higher partitioning of available water to evaporation. The parameter $d$ represents the baseflow and groundwater storage behaviour. The model uses rainfall and potential evaporation data as an input to simulate monthly streamflow. A schematic diagram for the dynamic water balance model is shown in Figure 4.3.

Details of the equations in figure 4.3 and model descriptions are presented in the Appendix B. This dynamic conceptual monthly water balance model has two storages: the root zone storage and groundwater storage and both act as linear storage reservoirs. State variables and fluxes are defined based in equation (4.4). Referring to equation (4.4), $w$ [-] is a model parameter ranging between 1 and $\infty$. For the purpose of model calibration Zhang $et$ $al.$ (2008) defined that $\alpha = 1 - \frac{1}{w}$, this implies that $\alpha$ [-] values vary between 0 and 1.

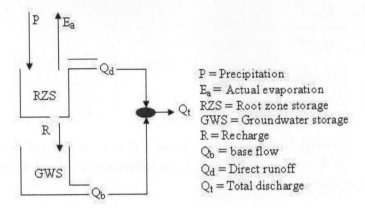

Figure 4.3: Schematic diagram of monthly water balance model structure; the unit of all variables for fluxes (P, $E_a$, $Q_d$, $Q_b$ and $Q_t$) is mm/month and mm for the storage parameters (RZS and GWS).

### 4.3.4 Parameter estimation, sensitivity and uncertainty assessment

In this study both manual and automatic calibration procedures have been done to estimate the best parameters set. All twenty catchments and the larger catchments of the Blue Nile at Kessie Bridge station and the Ethiopian-Sudanese border were calibrated on data from 1995-2000, and validated with data sets from 2001-2004 using split sample tests. For the manual calibration and validation, EXCEL spreadsheet model developed by Zhang et al. (2008) was used. However, in order to attain the optimal parameter set in the parameter space and to avoid the subjectivity of choosing parameters manually as well as for the parameter uncertainty assessment, in this study the EXCEL spreadsheet model was changed into a MATLAB code.

During the automatic calibration the Generalized Likelihood Uncertainty Estimation framework (GLUE) by Beven and Binley (1992) was employed to constrain model parameters using the Nash and Sutcliffe efficiency as likelihood measure. A threshold value of 0.7 for all catchments has been considered for behavioural models. Monte Carlo simulations employing 20,000 randomly parameter sets have been used to constrain the parameters in a feasible parameter space. Since the $E_{NS}$ was not improved beyond 20,000 simulations and the parameter uncertainty did not change for more simulations, the number of 20,000 simulations has been considered as sufficient for all catchments to constrain the parameters in the given parameters space. This is reasonable as was shown that the parameter space of much more complex models (e.g. a HBV model with 12 parameters) can be sampled sufficiently with 'only' 400,000 model runs (Uhlenbrook et al., 1999).

For the monthly model, we also considered a multi-objective optimization using two objective functions directed towards high flows and low flows. In this calibration process, all model parameters were simultaneously calibrated to minimize the objective function towards high flows and low flows. The concept of Pareto optimality is based on the notion of domination. It is used to solve the multi-objective optimization and derive Pareto-optimal parameter sets (Fenicia et al., 2007). All Pareto-optimal solutions are not dominated by the others. Mapping the Pareto-optimal solutions in a feasible parameter space produce Pareto-optimal front, this consists of more than one solution. Past research pointed out that calibration based on one single objective function often results in unrealistic representation of the hydrological system behaviour. This means, the information content of the data is not fully explained in a single objective function (Gupta et al., 1998; Fenicia et al., 2007). In a monthly model the following evaluation criteria were used in this study.

$$F_{HF} = \frac{\sum\limits_{i=1}^{n}(Q_{sim,i} - Q_{obs,i})^2}{\sum\limits_{i=1}^{n}(Q_{obs,i} - \overline{Q}_{obs})^2} \tag{4.7}$$

$$F_{LF} = \frac{\sum\limits_{i=1}^{n}[\ln(Q_{sim,i}) - \ln(Q_{obs,i})]^2}{\sum\limits_{i=1}^{n}[\ln(Q_{obs,i}) - \ln(\overline{Q}_{obs})]^2} \tag{4.8}$$

Where $Q_{sim,\,i}$ [mm month$^{-1}$] is the simulated streamflow at time $i$, $Q_{obs,\,i}$ [mm month$^{-1}$] is the observed streamflow at time $i$, $n$ is the number of time steps in the calibration period and the over-bar indicates the mean of observed streamflow. The objective function $F_{HF}$ was selected to minimize the errors during high flows (same formula as $E_{NS}$), and $F_{LF}$ uses logarithmic values of streamflow and improves the assessment of the low flows.

## 4.4 Results and discussions

### 4.4.1 Annual water balance

The annual water balance has been computed for the twenty catchments of the Upper Blue Nile basin with the assumption that evaporation can be estimated from water availability and atmospheric demand. Among the different Budyko's type curves, which were reported by Potter and Zhang (2009), Equation (4.4) has been applied to compute E/P for the twenty catchments. The selection of this equation was based on the simulation results from each catchment evaluated using Nash and Sutcliffe efficiency criteria. The predicted annual streamflow and evaporation results are given in Table 4.2. In most of these catchments, equation (4.4) couldn't predict the annual evaporation and stream flow adequately.

Table 4.2: Model performance using Equation (4.4) for annual runoff and annual evaporation during the period (1995-2004).

| Performance measure | Runoff | | Evaporation | |
|---|---|---|---|---|
| | min | max | min | max |
| $E_{NS}$ [-] | -2.37 | 0.84 | -0.39 | 0.73 |
| RMSE (mm a$^{-1}$) | 62.50 | 298.30 | 68.10 | 279.40 |

The poor accuracy of the prediction is related to the effect of neglecting soil water storage (assuming dS/dt = 0) and other error sources could influence the results as well, e.g. the uncertainty of catchment rainfall and the estimation of potential evaporation, seasonality of rainfall, and non-stationary conditions of the catchment itself (e.g. land use/land cover change). Furthermore, the year to year variability of rainfall depth could be the reason for the poor performance of the model. However, studying the effects of such factors on model simulation at annual time scale was not the objective of this study and needs further research.

Figure 4.4 illustrates the application of equation (4.4) to predict the regional long term annual mean water balance (1995-2004) of the twenty catchments. It can be seen that the aridity index varies from 0.7 to 1.5, and some catchments represent a semi-humid environment, when the aridity index less than 1.0 and others are drier with an aridity index greater than 1.0. Based on the aridity index, the Upper Blue Nile catchments can be classified as semi-arid to semi-humid and temperate.

Considering all twenty catchments, the regional mean annual water balance were adequately predicted by equation (4.4) with a single model parameter ($w = 1.8$) the model gave reasonable good performance with a Nash and Sutcliffe efficiency of ($E_{NS}$) 0.70 and a root mean squared error of 177 mm a$^{-1}$. The results of predicting the regional mean annual water balance using equation (4.4) in this study is in agreement with the research work in different parts of the world (e.g. Yang et al., 2007, Zhang et al., 2008; Potter and Zhang, 2009).

Furthermore, as it can be seen in Figure 4.4 that different groups of catchments follow a unique curve with an independent $w$ [-] value. Thus, the results depicts that catchment evaporation ratio varies from catchment to catchment. Based on evaporation ratio catchments are categorized into three groups. This suggests that each group has different characteristics. Study by Potter and Zhang, (2009) classified catchments in Australia based on rainfall regime as winter dominant, summer dominant, seasonal and non-seasonal catchments. Yang et al. (2007) also classified 108 arid to semi-arid catchments in China based on the patterns which the catchments follow on the

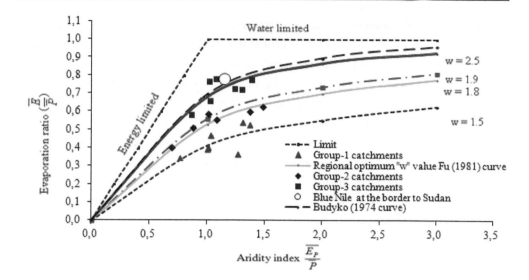

Figure 4.4: Mean annual evaporation ratio $\left(\dfrac{\overline{E}}{\overline{P}}\right)$ as a function of Aridity Index $\left(\dfrac{\overline{E_P}}{\overline{P}}\right)$ for different values of *w* using Fu (1981) curve.

Fu (1981) curve. Moreover, Zhang et al. (2004) categorized catchments in Australia based on vegetation cover as forested and grass land catchments with a higher parameter *w* values for forested catchments and lower values of the parameter *w* for grass covered catchments.

Table 4.3 presents goodness-of-fit statistics as recommended by Legates and McCabe (1999) for each group of catchments. The Nash and Sutcliffe efficiency ($E_{NS}$) and root mean squared error (RMSE) show that the long term average annual streamflow for each group of catchment was predicted adequately using equation 4.4.

Table 4.3: Average goodness-of-fit statistics Nash and Sutcliffe efficiency ($E_{NS}$) and root mean squared error (RMSE) for prediction of regional long term mean annual streamflow using Fu's curve.

|  | $E_{NS}$ (-) | RMSE (mm a$^{-1}$) | Calibrated parameter 'w' |
|---|---|---|---|
| All catchments | 0.70 | 177.51 | 1.8 |
| Group-1 catchments | 0.87 | 76.36 | 1.5 |
| Group-2 catchments | 0.97 | 57.22 | 1.9 |
| Group-3 catchments | 0.85 | 58.23 | 2.5 |

N.B: Group-1 catchments are: Gilgel Abay, Koga, Gumcra, Jedeb, Uke, Gilgel Beles and Beles.

Group-2 catchments are: Fetam, Dura, Guder, Muger, Temcha and Megech.

Group-3 catchments are: Neshi, Chemoga, Birr, Rib, Robi Gumero, Robi Jida and Didessa.

From the results of modelled individual catchments, it is noted that the calibrated parameter $w$ ranges between 1.4 and 3.6. This may suggest that the Upper Blue Nile catchments under consideration exhibit different catchment characteristics. Zhang et al. (2004) pointed out that smaller values of $w$ are associated with high rainfall intensity, seasonality, steep slope and lower plant available soil water storage capacity. However, it is difficult to represent these characteristics explicitly in a simple model (Zhang et al., 2004; Yang et al., 2007). Typically the result of the analysis from group-1 catchments in this study indicates that a larger fraction of precipitation becomes surface runoff, which results in lower evaporation ratios for this group compared to the other groups. Besides, the computed runoff coefficients for group-1 catchments ranges between 0.5-0.7 suggest that surface runoff was dominant in these catchments. The catchments in Group-2 have higher evaporation ratio and lower runoff coefficients (0.37-0.47) than Group-1. But Group-3 catchments have higher evaporation ratios, lower runoff potential and hence have lower values of the runoff coefficients (0.21-0.33). It is also noted that the parameter $w$ summarizes integrated catchment characteristics such as land cover, geology, soil properties and topography. It is not possible to fully explain the effects of $w$ for each group of catchment due to the lack of detail data of physiographic characteristics in the region.

### 4.4.2 Modelling streamflow on monthly timescale

Modelling at finer time scale (monthly and daily) requires the inclusion of soil moisture dynamics to accurately estimate the water balance. In a top-down modelling approach, model complexity has to be increased when deficiencies of the model structure in representing the catchment behaviour is encountered (Jothityangkoon et al., 2001; Atkinson et al., 2002; Montanari et al., 2006; Zhang et al., 2008). Therefore, a somewhat more complex model structure has been applied that is still very simple and has four parameters (see figure 4.3).

The monthly streamflows were calibrated for the period 1995-2000 and validated for the period 2001-2004. In the manual calibration and validation the Nash and Sutcliffe coefficient efficiency was used as leading performance measure. The main objective of calibration is finding the optimal parameter set that maximizes or minimizes the objective function for the intended purposes. In the parameter identification process, different parameter sets were sampled randomly from a priori feasible parameter space as shown in Table 4.4, which is in agreement with the literature (Zhang et al., 2008) and manual calibration in this study.

The dynamic water balance model was calibrated and validated for the twenty catchments and also at the two larger catchments with gauging stations at Kessie Bridge station (64,252 km$^2$) and at the Ethiopian-Sudanese border (173,686 km$^2$) to test the ability of the model at large spatial scale. During calibration Nash and Sutcliffe coefficients were obtained in the range of 0.52-0.95 (dominated by high flows) and 0.33-0.93 using logarithmic discharge values when calculating $E_{NS}$ (dominated by low flows).

44

Table 4.4: Ranges of parameter values for the catchments modelled in the Upper Blue Nile basin based on manual calibration and (Zhang et al., 2008).

| Catchment | Lower / upper bound | | | |
|---|---|---|---|---|
| | $S_{max}$ [mm] | $\alpha_1$ [-] | $\alpha_2$ [-] | d [1/month] |
| Gilgel Abay, Koga, Birr, Fetam, Neshi | 100 - 600 | 0 - 1 | 0 - 1 | 0 - 1 |
| Dura, Gilgel Beles, Gumera, Megech, Rib, Robigumero, Robijida, Didessa | 100 - 600 | 0.1 - 0.75 | 0.1 - 0.75 | 0 - 1 |
| Chemoga, Beles, Guder | 100 - 600 | 0.1 - 0.85 | 0.1 - 0.85 | 0 - 1 |
| Muger, Temcha, Uke, Jedeb | 100 - 600 | 0 - 0.9 | 0 - 0.8 | 0 - 1 |

Similarly, during the validation period Nash and Sutcliffe efficiencies were obtained in the range of 0.55-0.95 during high flows and 0.12-0.91 during low flows. The model results reveal that during calibration, the model gave reasonable results in most of the catchments including the simulation at larger scale at Kessie Bridge station ($E_{NS}$ = 0.95) and at the Ethiopian-Sudanese border ($E_{NS}$ = 0.93). However, during the validation period in some catchments the low flows were not captured well by the model. Though a lot of uncertainties in model structure and model parameters were common in hydrological models, it is speculated that the likely reason for poor efficiency values in some catchments were more related to the uncertainties in the input data sets. Figure 4.5 shows observed and predicted streamflows using automatic calibration for selected meso-scale catchments and the larger scale results at Kessie Bridge station and Ethiopian-Sudanese border.

It is clearly demonstrated that the parameter values differ and the performance of the model improved using a GLUE framework. The parameter $\alpha_1$ in majority of the catchments shows that the rainfall amount retained by the catchments is not significant, thus fast runoff generation process are dominant in the studied catchments. This high responsiveness is also in line with field observations where the formation of surface runoff (and significant soil erosion) can be observed. The values of evaporation efficiency parameter $\alpha_2$ are higher in some catchments which imply higher partitioning of available water into evaporation. The higher evaporation efficiency parameter in these catchments reveals that the catchments were relatively having high forest cover and evaporation was dominant and less surface runoff was generated in these catchments. The dotty plots used to map the parameter value and their objective function values as a means of assessing the identifiability of parameters are shown in Figure 4.6. The optimal parameter values obtained using GLUE framework together with parameters obtained manually are presented in Table 4.5.

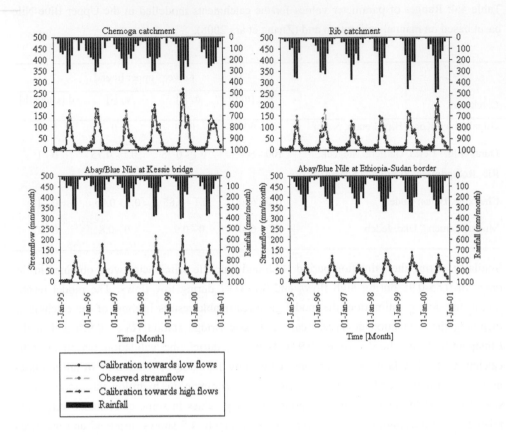

Figure 4.5: Observed and simulated streamflows during calibration period (1995-2000) for selected   catchments.

Table 4.5: Comparison of automatic (GLUE) single objective and manually calibrated parameters and $E_{NS}$ values of the twenty selected Upper Blue Nile catchments during the period 1995-2000.

| Catchment | Optimized parameters (GLUE) | | | | | Manually calibrated parameters | | | | |
|---|---|---|---|---|---|---|---|---|---|---|
| | $S_{max}$ [mm] | $\alpha_1$ [-] | $\alpha_2$ [-] | d[1/T] | $E_{NS}$ [-] | $S_{max}$ [mm] | $\alpha_1$ [-] | $\alpha_2$ [-] | d [1/T] | $E_{NS}$ [-] |
| Beles | 538.09 | 0.46 | 0.53 | 0.96 | 0.71 | 365.00 | 0.50 | 0.52 | 0.85 | 0.70 |
| Birr | 253.62 | 0.76 | 0.92 | 0.87 | 0.93 | 330.00 | 0.57 | 0.62 | 0.91 | 0.82 |
| Chemoga | 216.40 | 0.66 | 0.81 | 0.89 | 0.93 | 260.00 | 0.60 | 0.63 | 0.90 | 0.90 |
| Didessa | 590.02 | 0.58 | 0.85 | 0.48 | 0.52 | 520.00 | 0.64 | 0.86 | 0.23 | 0.52 |
| Dura | 331.56 | 0.65 | 0.63 | 0.94 | 0.92 | 360.00 | 0.63 | 0.60 | 0.86 | 0.91 |
| Fetam | 420.89 | 0.56 | 0.50 | 0.99 | 0.89 | 280.00 | 0.52 | 0.67 | 0.35 | 0.81 |
| Gilgel Abay | 349.30 | 0.71 | 0.88 | 0.12 | 0.82 | 390.00 | 0.60 | 0.86 | 0.60 | 0.81 |
| Gilgel Beles | 280.03 | 0.57 | 0.35 | 0.98 | 0.91 | 280.00 | 0.55 | 0.47 | 0.90 | 0.89 |
| Guder | 306.01 | 0.83 | 0.45 | 0.94 | 0.72 | 390.00 | 0.67 | 0.41 | 0.85 | 0.71 |
| Gumera | 230.82 | 0.68 | 0.39 | 1.00 | 0.73 | 350.00 | 0.47 | 0.41 | 0.89 | 0.72 |
| Jedeb | 315.96 | 0.58 | 0.43 | 0.93 | 0.79 | 260.00 | 0.58 | 0.41 | 0.75 | 0.78 |
| Koga | 200.14 | 0.64 | 0.46 | 0.51 | 0.79 | 240.00 | 0.60 | 0.48 | 0.60 | 0.79 |
| Megech | 313.58 | 0.55 | 0.61 | 0.98 | 0.81 | 390.00 | 0.55 | 0.62 | 0.78 | 0.78 |
| Muger | 190.52 | 0.79 | 0.60 | 0.97 | 0.88 | 210.00 | 0.63 | 0.60 | 0.85 | 0.84 |
| Neshi | 397.90 | 0.79 | 0.82 | 0.93 | 0.80 | 380.00 | 0.66 | 0.64 | 0.75 | 0.75 |
| Rib | 558.43 | 0.52 | 0.72 | 0.99 | 0.82 | 370.00 | 0.63 | 0.73 | 0.60 | 0.76 |
| Robigumero | 512.88 | 0.53 | 0.73 | 0.02 | 0.74 | 500.00 | 0.50 | 0.75 | 0.65 | 0.73 |
| Robijida | 243.61 | 0.69 | 0.75 | 0.02 | 0.91 | 420.00 | 0.52 | 0.78 | 0.90 | 0.86 |
| Temcha | 190.80 | 0.66 | 0.63 | 0.99 | 0.83 | 200.00 | 0.62 | 0.60 | 0.80 | 0.82 |
| Uke | 441.60 | 0.70 | 0.33 | 0.74 | 0.82 | 430.00 | 0.67 | 0.33 | 0.70 | 0.82 |
| Blue Nile at Kessie Bridge | 268.25 | 0.70 | 0.77 | 0.53 | 0.95 | | | | | |
| Upper Blue Nile at the border to Sudan | 439.37 | 0.76 | 0.74 | 0.39 | 0.93 | | | | | |

It can be seen that most of the parameters in the Upper Blue Nile catchments are reasonably well identifiable; however, the recession constant $d$ exhibit poor identifiability in the majority of the catchments. It is speculated that parameter uncertainty and model structural errors could be the reason for the poorly identifiable groundwater parameter. Furthermore, the model performances in the objective function space of Pareto-optimal fronts resulting from the monthly water balance model were investigated (see Figure 4.7).

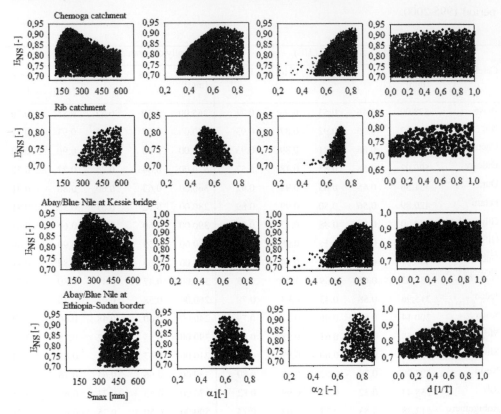

Figure 4.6: GLUE dotty plots of selected meso-scale catchments and Blue Nile at larger scale at Kessie Bridge and Ethiopian -Sudanese border.

From multi-objective optimization point of view, the Pareto-optimal solutions are all equally important to achieve a better model simulation. The Pareto based approach is also important to compare different model structures in such a way that model improvement can be attained as the Pareto-optimal front progressively moves towards the origin of the objective function space (Fenicia et al., 2007; Wang et al., 2007a). From figure 4.7 it can be noticed that for different catchments the Pareto-optimal set of solutions approach to the origin differently. It is demonstrated that the model structure performs better at the larger scale than for the meso-scale catchments. As the objective function values get closer to the origin, the chosen model structure represents the hydrologic system better.

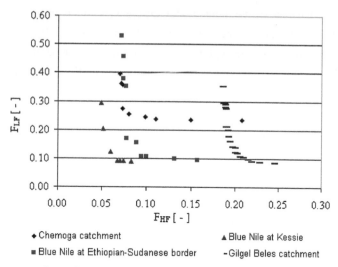

Figure 4.7: Pareto-optimal fronts of parameter sets at different meso-scale catchments and Blue Nile at the larger scale based on the selected objective functions [low flows ($F_{LF}$), high flows ($F_{HF}$)].

## 4.5 Conclusions

The Abay/Upper Blue Nile catchment water balance has been analyzed at different temporal and spatial scales using Budyko's framework. The analysis included water balance at mean annual, annual and monthly time scales for meso to large scale catchments. Budyko-type curve (Fu, 1981) was applied to explore the first order control based on available water and energy over mean annual and annual time scales. The results demonstrated that predictions are not good in the majority of the catchments at annual time scale. This implies that at annual scale the water balance is not dominated only by precipitation and potential evaporation. Thus, increased model complexity to monthly time scale is necessary for a realistic simulation of the catchment water balance by including the effects of soil moisture dynamics. Parameters were identified using the Generalized Likelihood Uncertainty Estimation (GLUE) framework in addition to manual calibration and the results showed that most of the parameters are identifiable and the model is capable of simulating the observed streamflow quite well. The applicability of this model was tested earlier in 250 catchments in Australia with different rainfall regimes across different geographical regions and results were encouraging (Zhang et al., 2008; Fang et al., 2009). Similarly, in the Upper Blue Nile case, the model performs well in simulating the monthly streamflow of the twenty investigated catchments.

With only four parameters the simple model has the advantage of minimal equifinality. Despite the uncertainties in input data, parameters and model structure, the model gives reasonable results for the Upper Blue Nile catchments. However, it is suggested that on annual time scale the reasons for poor model efficiencies in majority of the catchments, which followed distinct Budyko-type curves, needs further research. It is recommended that future work should focus on the regionalization of the optimal parameter sets from the monthly model presented in this study for prediction of streamflow in ungauged catchments in the Upper Blue Nile basin.

# CHAPTER 5

Hydrologic responses to land cover change: the case of Jedcb meso-scale catchmcnt, Abay/Upper Blue Nile basin, Ethiopia[1]

In the previous chapters, trends in hydro-climatic variables and water balance study were presented. Some of the causes of streamflow changes were studied in more detail by investigating the effects of land use change on hydrologic responses in the Jedeb meso-scale agricultural dominated catchment. Land use change is caused by a multiple of interacting factors. Among these factors, population pressure, intensive agriculture, over grazing, and deforestations are considered to be the primary factors in changing the river flow regime. Moreover, their integrated effects are noticeable by investigating the streamflow at catchment outlet. Jedeb meso-scale catchment is one of the headwater of the Abay/Upper Blue Nile basin. It is most important in terms of the water sources of the main Abay/Blue Nile. However, intensive agricultural practices in unfavourable land slope, soil erosion, and land degradation are the major problems, which likely contributing to the change in responses of the catchment. Therefore, this chapter provides the study of land use change impacts on the hydrologic responses of Jedeb catchment through use of statistical tests, and conceptual hydrological model.

## 5.1 Introduction

Quantifying the effects of land use change on streamflow has recently been given much attention by the scientific communities (e.g. Jiang et al., 2011; Yang et al., 2012). Land use change is caused by a multiple of interacting factors of the coupled human and environment systems (Lambin et al., 2003).

Land use change is the main causes for soil degradation and could significantly change the streamflow availability (Tolba et al., 1992). This is attributed to different anthropogenic activities, e.g. intensive land cultivation, deforestation, overgrazing, urbanization, afforestation and reforestation. The heterogeneity of topography, land use, soil and geology, on the one hand and limitation of hydro-meteorological data on the other hand are considered to be the major scientific challenges often preclude to study land use change impact on

---

[1] *This chapter is based on*: Tekleab, S., Mohamed, Y., Uhlenbrook, S., and Wenninger, J. (2014). Hydrologic responses to land cover change: the case of Jedeb meso-scale catchment, Abay/Upper Blue Nile basin, Ethiopia. Hyrol. Process. J., doi: 10.1002/hyp.9998, 28, 5149-5161.

hydrology (DeFries and Eshleman, 2004). In fact, the underlying mechanisms, which underpin the impact of land use change on streamflow is not yet fully understood (Wang et al., 2007b).

The literature shows many studies in different geographical regions investigating the effects of land use/land cover change (LULC) change on streamflow using a small experimental paired catchment approach (e.g. Bosch and Hewlett, 1982; Zhang et al., 2001a; Brown et al., 2005). Similar studies using hydrological models at a catchment scale are documented in e.g. (Siriwardena, et al., 2006; and Seibert and McDonnell. 2010). However, the research outputs from different geographical regions indicated different and sometimes contradicting results (Rientjes et al., 2011b; Costa et al., 2003; Siriwardena, et al., 2006).

In Ethiopia, Bewket and Sterk (2005) reported a decrease of the dry season flows in Chemoga catchment (adjacent to the Jedeb catchment), attributed to clearance of natural vegetation cover, increased agricultural expansion, and growing areas of eucalyptus plantation. Most recently, Rientjes et al., (2011b) showed that both LULC change and seasonal distribution of rainfall were the causes of streamflows changes of the Gilgel Abay River (source of the Abay/Upper Blue Nile river).

Land degradation is the most tempting problem threatening the subsistence agriculture in most of the Northern Ethiopian highland. A study by Hurni (1993) showed that the estimated soil loss rate from crop land in the Northern Ethiopian highland was 42 t/ha/year, and it reaches up to 300 t/ha/year for the individual fields. However, empirical studies about the cause and effects of those changes on water resources are very limited. The existing results from researches were mainly focused on land cover detection analyses (e.g. Zeleke and Hurni, 2001; Bewket and Sterk, 2005; Hurni et al., 2005; Teferi et al., 2010; Rientjes et al. 2011b).

Quantifying the impact of land use change on streamflow is of key importance for transboundary water management at a river basin scale (e.g. the Nile basin). The literature showed extremely limited studies to quantify such externalities in the Abay/Upper Blue Nile basin (Teferi et al., 2010; Tekleab et al., 2011; Gebremichael et al., 2012; Tessema et al. 2010). Nonetheless, those studies have used aggregated time steps (e.g. monthly data), that miss the dynamics of daily flow variability.

Therefore, the main objective of this study is to quantify the effects of LULC change on Jedeb streamflows using two different methods: (i) Assessing catchment response by quantifying statistical trends of the daily streamflow data, (ii) Examining the streamflow regime in the Jedeb catchment using a monthly hydrological model for the period 1973-2010.

## 5.2. Study area

The Jedeb river is a tributary of the Abay/Upper Blue Nile river, located south of Lake Tana, see Figure 5.1. It originates from Choke Mountains at an elevation of 4000 m. a.s.l. The drainage area upstream of the stream gauging station covers 296 km$^2$. The long term catchment average annual temperature over the period of 1973-2010 is about 16°C. The precipitation ranges between 1326 to 1434 mm a$^{-1}$ in the lower and upper part of the catchment respectively. The long term mean streamflow is about 5.5 m$^3$ s$^{-1}$ (see Table 5.1). Due to scarce weather data, potential evaporation was computed using the Hargreaves method, and amounts to 1374 mm a$^{-1}$ (Hargreaves and Samani, 1982). The soil types in the catchment are Haplic Alisols, which are deep, silty clay soils; Haplic Luvisols, which are well drained soils with clay to silty clay texture and Eutric Leptosols, as moderately deep soils with a clay loam to clay texture (BCEOM, 1998a).

The main geological formation in the study area is Termaber basalts which are parts of the Tertiary Shield group. Termaber basalts are underlain by the Blue Nile basalts, which are parts of Ashangi group and it is underlain by the Mesozoic Adigrat sandstones (BCEOM, 1998b).

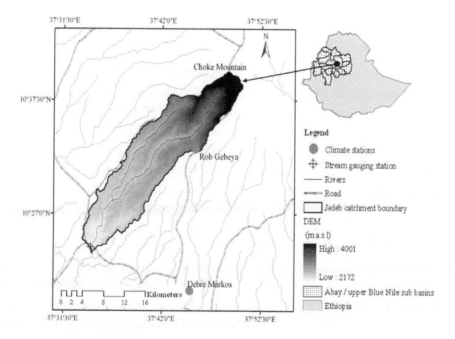

Figure 5.1: Location and topography of study area.

## 5.3 Data sources

### 5.3.1 Hydro-meteorological data

The streamflow data set is based on manual water level measurements (daily at 06:00 a.m. and 06:00 p.m.) at 'Yewla' gauging station for the period 1973-2010, and was collected from the Ethiopia Ministry of Water Resources and Energy. Daily precipitation and temperature data at Debre Markos and Rob Gebeya stations were obtained from the Ethiopian National Meteorological Agency. Table 5.1 shows the periods of the collected data records.

Table 5.1: Description of daily hydro-meteorological data availability in the Jedeb catchment.

| Station | Period of record | Long term annual mean | Coefficient of variation | Remark |
|---------|------------------|-----------------------|--------------------------|--------|
| Debre Markos[a] | 1954-2010 | 1326 (mm a$^{-1}$) | 12% | <1% missing data |
| Rob Gebeya[a] | 1989-2010 | 1434 (mm a$^{-1}$) | 14% | No missing data |
| Debre Markos[b] | 1973-2010 | 16.3 ($^\circ$C) | 3.5% | 2% missing data |
| Jedeb [c] | 1973-2010 | 593 (mm a$^{-1}$) | 24% | 5% missing data * |

*Data record for the years 1978, 1991, 1992, 1993, 1994, 2000, 2001 and 2002 were excluded from the analysis. The superscript a, b, and c represent the precipitation, temperature and streamflow respectively.

As a first step, suspicious records of the hydro-meteorological data were omitted. The consistency of precipitation records at Debre Markos and Rob Gebeya were checked using a double mass analysis against neighbouring precipitation stations at Anjeni, Lumame and Dejen. The precipitation data at both stations were consistent. Similarly the discharge series was checked against outliers using a comparison with the neighbouring hydrograph of the Chemoga stream gauging station. Next to that a new rating curve equation for the years 2009-2010 was developed using regressions model. Suspicious discharge data, e.g., with annual runoff values being larger than annual rainfall values have been neglected in our analysis.

### 5.3.2 Land use land cover data

Land use/land cover information of the Jedeb catchment was analysed recently by (Teferi et al., 2013). They used aerial photographs of 1957, Landsat imageries of 1972 (MSS), 1986 (TM), 1994 (TM), and 2009 (TM) to quantify the long term LULC in the Jedeb catchment. Ten land use / land cover classes were identified in the Jedeb catchment, given in Table 5.2.

Table 5.2: The major land use / cover types and their percentage within Jedeb catchment over the year 1957, 1972, 1986, 1994 and 2009 (Source: Teferi et al., 2013).

| Land use / cover | 1957 | 1972 | 1986 | 1994 | 2009 |
| --- | --- | --- | --- | --- | --- |
| | % | % | % | % | % |
| GL | 20 | 20 | 17 | 14 | 15 |
| AGL | 6 | 6 | 5 | 5 | 5 |
| SHB | 7 | 3 | 2 | 2 | 2 |
| CL | 53 | 62 | 67 | 69 | 70 |
| RF | 4 | 2 | 2 | 2 | 1 |
| WL | 4 | 3 | 2 | 2 | 1 |
| PF | 0 | 0 | 3 | 3 | 3 |
| ML | 4 | 3 | 1 | 2 | 2 |
| BL | 1 | 0 | 1 | 1 | 1 |
| EF | 1 | 1 | 0 | 0 | 0 |

Where: GL is grass land, AGL: afroalpine grass land, SHB: shrub and bush, CL: cultivated land, RF: riverine forest, WL: wood land, PF: plantation forest, ML: marsh land, BL: bare land and EF is Ericaceous Forest.

## 5.4. Methodology

Two methodological approaches were used to assess the impact of LULC change on streamflow. First, evaluating variability of streamflow magnitude, frequency and duration of the daily flow time series using statistical parameters such as flashiness index, flow reversal and rise and fall rate of the hydrograph. Secondly, by analysing the parameters of a monthly time step hydrologic model to detect if a LULC change has occurred. The details for the methodological approach are given in the subsequent sections.

### 5.4.1 Flow variability analysis

Flow variability in the Jedeb catchment has been analysed following the approach of Archer (2000, 2007). Basically it is based on analysing the frequency and duration of pulses above/below certain flow thresholds. A discharge pulse is defined as an event occurs above or below a certain threshold (see Figure 5.2 top panel.). Different thresholds of 0.1M to 80M have been selected. The flow thresholds for representing pulses were multiples of the median discharge (M) of the Jedeb river 1.21 $m^3 s^{-1}$ computed over the period 1973-2010. The annual number and duration of discharge pulses has been computed from the daily discharge series between 1973 and 2010. The computation has been done by counting the number of pulses in each hydrological year for each threshold, to create a time series for the given parameters.

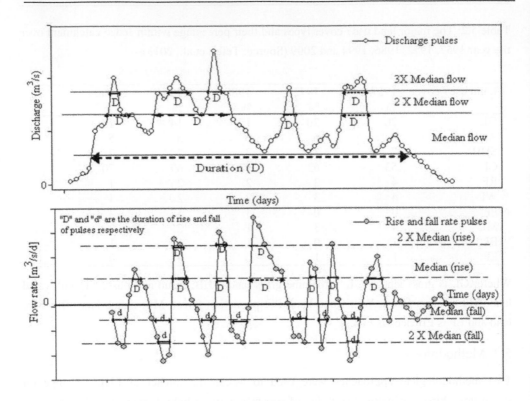

Figure 5.2: Definition sketch for pulse number and pulse duration: top panel represent the flow and lower panel represent flow rates in a given year (axes are not to scale).

Additionally, the flashiness index called the Richard Baker index (R-B index) of pulses for each year has been computed to assess frequency and rapidity in streamflow (Baker et al., 2004).

$$R - B\ Index = \frac{\sum_{i=1}^{n}|q_i - q_{i-1}|}{\sum_{i=1}^{n} q_i} \tag{5.1}$$

Where, $q_i$ and $q_{i-1}$ ($m^3\ s^{-1}$) is the daily discharge at time step i, and i-1, respectively, n is the number of observations. The index has low inter-annual variability and is relatively better to detect trends in flow regimes (Baker et al., 2004).

One indicator of hydrological alteration is the reversal flow. It is the change in the direction of the daily discharge (rise versus fall). The average rise and fall rate of the hydrograph were computed using hydrologic alteration (IHA) software (e.g. Richter et al., 1996). The rise rate is the positive difference between consecutive daily discharges, while fall rate is the negative difference. The median rates of the rise and fall for the whole period are computed as 0.24

$m^3/s/d$ and -0.85 $m^3/s/d$, respectively. Analogous to the median flow (M), multiples of median rise and fall rate thresholds were defined (see figure 5.2 lower panel). Threshold level is selected arbitrary above or below the median flow and the median of rise and fall rate of the hydrograph.

The statistical significance of trends was examined using Spearman's Rank correlation non parametric test (e.g. Masih et al., 2010). Two series are compared related to the rank of the data. $K_{xi}$ is the rank of the data as it was measured. $K_{yi}$ is the series of the rank of the same data in ascending or descending order. The null hypothesis reads: no trend between the order in which the data were observed and the increase or decrease in magnitude of those data. The alternative hypothesis is $X_i$ increase or decrease with i, indicating that a trend exists.

The Spearman coefficient of rank correlation $R_{sp}$ is then defined as:

$$R_{sp} = 1 - \frac{6 * \sum D_i^2}{n * (n^2 - 1)}$$

(5.2)

$$D_i = K_{xi} - K_{yi}$$

(5.3)

$$t_t = R_{sp} * \left(\frac{n-2}{1-R_{sp}^2}\right)^{0.5}$$

(5.4)

The test statistic $t_t$ has student's t-distribution and n is the number of elements in the samples.

### 5.4.2 Monthly Flow duration curve

Monthly flow duration curve of the observed discharge time series is computed for the different time period to evaluate flow regime change in relation to the LULC condition in 1970s, 1980s, 1990s and 2000s. The flow duration curve describes the streamflow against the percentage of time in which the streamflow is equalled or exceeded. It is a useful parameter for representing streamflow variability (Shao et al., 2009).

### 5.4.3 Hydrological simulation model

The monthly water balance model developed by Zhang et al. (2008) has been used to assess the effects of LULC on the runoff from Jedeb catchment. This model has given good results when applied to the Abay/Upper Blue Nile basin (Tekleab et al., 2011). The model structure is shown in Figure 4.3 (see Chapter 4), with four parameters: (i) $\alpha_1$ [-] represents catchment rainfall retention. An increase in $\alpha_1$ implies higher rainfall retention and less direct runoff. (ii) $S_{max}$ [mm] is the maximum soil water storage in the root zone. It relates to soil and vegetation characteristics of the catchment. (iii) $\alpha_2$ [-] relates to the evaporation efficiency. A higher value of $\alpha_2$ implies higher partitioning of available water to evaporation. (iv) d [1/month] represents the baseflow and groundwater storage behaviour.

A uniform random sample of 20,000 Monte Carlo run has been used to identify model parameters (Tekleab et al., 2011). Details of parameters range and equation describing the model structure can be found in appendix 4 of this thesis or (Zhang et al., 2008; Tekleab et al., 2011). The two performance measures Nash and Sutcliffe coefficient of efficiency ($E_{NS}$), Eq. (5.5), and root mean squared error (RMSE), Eq. (5.6), were used to evaluate the calibrated parameters over successive calibration periods.

$$E_{NS} = (1) - \frac{\sum_{i=1}^{n}(Q_{sim,i} - Q_{obs,i})^2}{\sum_{i=1}^{n}(Q_{obs,i} - \overline{Q}_{obs,i})^2}$$

(5.5)

$$RMSE = \sqrt{1/n * \sum_{i=1}^{n}(Q_{obs,i} - Q_{sim,i})^2}$$

(5.6)

Where $Q_{sim,i}$ is the simulated streamflow at time $i$ [mm/month], $Q_{obs,i}$ is the observed streamflow at time $i$ [mm/month], $n$ is the number of time steps, and the over-bar indicates the mean of observed streamflow.

Optimal model parameters might differ when a model is calibrated for different time periods due to LULC change (Seibert and McDonnell, 2010). Therefore, the model has been calibrated/validated separately for four time periods, viz: 1970s, 1980s, 1990s and 2000s, as given in figure 5.3 To infer LULC change, the results of calibrated model parameters over different time periods are interpreted in relation with the results of the flow variability approach in section 5.4.1. The basic assumption here is that the model results evaluate the hypothesis that increasing/decreasing in flashiness of the Jedeb catchment is attributed mainly to a decrease / increase in soil moisture retention over time.

For comparison of model parameters and the physical interpretation of catchment processes for the different time periods, the study has also examined the posterior cumulative distribution of 100 (arbitrarily selected) best performing parameters from the optimal parameter sets. This avoids the selection of single optimal model, which can complicate the interpretation of changes in parameter values due to parameter interaction (Seibert and McDonnell, 2010). The plot of cumulative distribution explains how the different optimal parameters over varying time periods exhibit the response of the catchment. Consequently, the changes of cumulative distribution of parameter values for different calibration periods might be related to LULC changes in the catchment.

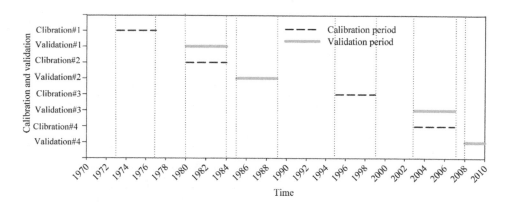

Figure 5.3: Different time periods indicating calibration and validation of the monthly water balance model.

## 5.5 Results and discussion

### 5.5.1 Flow variability analysis

The changes in daily flow dynamics can be detected from the results of flow variability analysis over the period 1973-2010. The discharge time series for the analysis is shown in Figure 5.4 and the result of the statistical analysis of pulses is given in Table 5.3.

The results presented in Table 5.3 show that the annul number of pulses decreases for the flow range above 15 to 25M, while increases for very high thresholds >70M. The duration of pulses decreases for lower thresholds (0.1 to 20M), and increases for very high thresholds (70 and 80M). This could mean that the duration of the flow pulses exhibits statistically significant decreasing trend for a wide spectrum of flows (0.1 to 20M). The analysis of the number of pulses for different thresholds representing low flows occurred in each of the four time periods is shown in Figure 5.5. For the higher thresholds more pulses for median flow is noticeable in the later period 2000's compared to earlier periods. On the other hand, pulses below certain threshold values, representing the low flow regimes, are increasing in the later periods (Figure 5.5). This implies that low flows are decreasing towards the later periods.

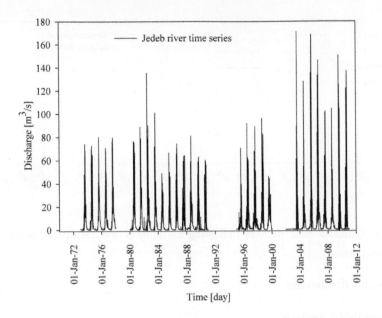

Figure 5.4: Time series discharge data of Jedeb river 1973-2010

Table 5.3: t statistics from Spearman's rank correlation test for varying flow thresholds during the period 1973 to 2010. Bold numbers designates statistically significant trends at 5% significance level. Positive and negative signs indicate the direction of trend.

| Thresholds | 0.1M | 10M | 15M | 20M | 25M | 30M | 40M | 70M | 80M |
|---|---|---|---|---|---|---|---|---|---|
| Discharge (m³/s) | 0.12 | 12.1 | 18.2 | 24.2 | 30.3 | 36.3 | 48.4 | 84.7 | 96.8 |
| Annual number of pulses (t-statistics) | 0.97 | -1.77 | **-3.46** | **-2.54** | **-2.09** | -0.97 | 1.01 | **4.92** | **5.28** |
| Annual duration of pulses (t-statistics) | **-2.63** | **-2.63** | **-3.44** | **-2.55** | -1.58 | -0.97 | 0.59 | **2.11** | **2.25** |

This means that the magnitudes of the low flows below a certain thresholds in the earlier periods are greater than the later periods. These in turn reflects or speculates that the catchment becoming more flashy to the rainfall input and decrease the amount of water infiltrating into the groundwater during recharge periods.

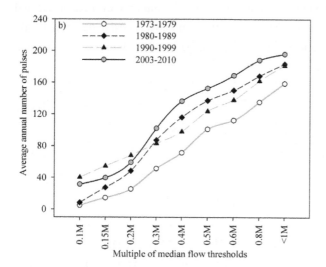

Figure 5.5: Average annual frequency of pulses exists in a time period for multiple of median flow thresholds (low flow regime).

### 5.5.2 Indicators of hydrologic alteration

### The rise and fall rate of the flow hydrograph

The number of pulses related to rise and fall rate of the hydrograph provides good insights about how rapidly/slowly is the respond of the flow hydrograph to the given climatic input (basically rainfall characteristics) and the LULC condition of the catchment.

Table 5.4 presents the annual number and duration of pulses for rise and fall rates of the hydrograph. The number of rise and fall pulses depicts statistically significant increasing trends, particularly for the high thresholds. While, duration of pulses show a consistently decreasing trends for rise rate and it is statistically significant at low threshold values. This indicates that the flow hydrograph is flashy and the duration of rise is decreasing over time, though the duration is not statistically significant for thresholds >25M.

Table 5.4: t- Statistics from Spearman's rank correlation test for multiple of median rise and fall rate thresholds over the period 1973-2010. Bold numbers designate statistically significant trends at 5% significance level. Positive and negative signs indicate the direction of trend and 'na' means not available.

| Thresholds | 0.1M | 10M | 25M | 50M | 75M | 100M |
|---|---|---|---|---|---|---|
| Rise rate (m$^3$/s/d) | 0.02 | 2.34 | 5.85 | 11.70 | 17.75 | 23.4 |
| Annual number of pulses for rise rate (t-statistics) | -1.05 | 1.95 | 1.96 | **2.08** | **2.50** | **2.32** |
| Annual duration of pulse for rise rate (t-statistics) | **-10.7** | **-7.71** | -1.89 | -0.13 | -0.37 | -0.04 |

| Thresholds | <0.1M | <0.5M | <10M | <25M | <50M | <75M |
|---|---|---|---|---|---|---|
| Fall rate (m$^3$/s/d) | -0.09 | -0.43 | -8.5 | -21.25 | -42.5 | -63.75 |
| Annual number of pulses for fall rate (t-statistics) | 0.54 | 1.31 | **2.66** | **2.62** | **4.03** | **5.40** |
| Annual duration of pulse for fall rate (t-statistics) | **-2.08** | -1.84 | -0.93 | -0.94 | -1.93 | na |

The results of the annual number of pulses for different decades also indicate an increase in the later periods 1990s and 2000s. The increased annual pulse number for rise and fall rate in later periods suggests faster rate of flows and the reduction of the discharge rate is more rapid than in earlier periods. This implying that the increased intensification of agricultural practices with poor land management could contribute to fast changing flow hydrographs and possibly lead to an increased gully formation. The analysis of average annual rise and fall rate of the hydrograph also suggests a statically significant trend evaluated using the Spearman's correlation test (t = 3.45 and -2.22), respectively (Figure 5.7c).

**Flashyness Index:**

The average daily flow data from 1973-2010 was used to analyse the flashiness of the Jedeb catchment. Figure 5.7a illustrates the pattern of the mean annual flashiness index over the period 1973-2010, which shows a statistically significant increasing trend (student t-statistics = 3.62 using the Spearman's correlation test at 5% level).

The occurrence of a large number of pulses from the analyses of median flows and the rise and fall rate pulse numbers demonstrate the increasing trend of R-B index. This suggest that the catchment is becoming more responsive to the rainfall input over time and more apparent in the later periods. Tekleab et al. (2013) showed that the rainfall trend from Debre Markos climate station in close proximity of the Jedeb catchment reveal no evidence of increasing or decreasing trends on mean annual, monthly or seasonal time scales. Moreover, the same results have been obtained by evaluating the statistical significance of frequency of wet days above different threshold values which can be an indicator of rainfall intensity (Orr and Carling, 2006). The results confirm no evidence of statistical trends of frequency of wet days above a threshold values at Debre Markos and Rob Gebeya climate stations around the Jedeb catchment. However, the average annual temperature from the climate station at Debre Markos depicts statistically increasing trend evaluated at 5% significance level. Thus, land use change or agricultural management practices might have more pronounced effects than the precipitation change, which responsible for the hydrological change in the catchment.

A recent study in the Jedeb catchment by Teferi et al. (2013) reported that, from the total catchment area in 1950s, the cultivated land coverage has increased from 53% to 70% in 2000s, see Table 5.2 for details. There is a 17% increase in crop land between 1957 and 2009. It is plausible to say that the flashiness in Jedeb catchment is more likely linked to the land use practices, given the fact that rainfall pattern remains the same. For example, most of the population in the catchment is living under the dominant slope class between 15-30%, in which intensive cultivation and deforestation pervasive in this slope category over the last five decades (Teferi et al., 2013). Based on the study of Hurni (1988), if the slope class in cultivated land is between 5-15% then, the cultivated land can be vulnerable to erosion risk. Therefore, it can be deduced that the farming practices in this dominant cultivated land slope class in the Jedeb catchment could be one of the causes for alteration of the hydrograph flow characteristics.

Moreover, during the field campaign in July 2009-Aug. 2012, it has been seen that farmers are traditionally draining the excess water during the runoff events from their farm using farm ditch. Farmers made the ditches along the slope, which is suitable for the initiation of the gully formation. From field observations in the Jedeb catchment, and inquiry of the local inhabitants, gullies have formed in late 1990s and the most devastating were in the recent years (see fig.5.6). Consequently, the gullies are accelerating the river flow at faster rate, which lead to increased rapidity of Jedeb river flow.

Figure 5.6: Soil erosion (left) and farming practice in hillslope area (right) in the Jedeb catchment (Photo taken by the author of this thesis).

**Flow reversal:**

The hydrograph reversal describes the change in the direction of rise and fall of the hydrograph. As it is shown from Figure 5.7b that reversal count in a year over the past four decades indicate statistically significant increasing trend based on Spearman's rank correlation test (t = 6.41). Moreover, flow variability indicator such as the 1day annual maximum and minimum flow in the Jedeb catchment showed statistically significant increasing trend for high flows and a decreasing direction of trend for low flows (Tekleab et al., 2013).

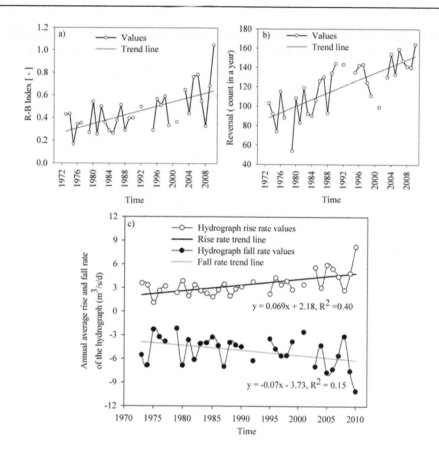

Figure 5.7: Flow variability indicators: a) Flashiness index (R-B Index), b) Reversal flow, which is change in the direction of the flow and, c) rise and fall rate of the hydrograph.

The flow variability indicators analyzed above, indicated a statistically significant change of the flow hydrograph over the past four decades. Nevertheless, linking these changes to LULC change is not possible using only statistical approaches. Therefore, a combined use of statistical methods and a hydrologic model could be a viable approach to assess the effects of LULC change on catchment runoff (Lorup et al., 1998). The next section discusses the application of a conceptual hydrological model to evaluate the influence of land use change on flow hydrograph. Basically to test the hypothesis that, rapid catchment response was related to lower rainfall retention by the soil.

### 5.5.3 The hydrological simulation model

#### Calibration and validation

The monthly time step model has been calibrated and validated for four time periods. The hypothesis is that an increase of the flashiness is attributed to a reduced infiltration capacity of the soil. Figure 5.5 shows the calibration/validation time periods. Figure 5.8 presents the results of the hydrographs for the four periods during calibration and validation. The model results depict that the performance measure Nash-Sutcliffe efficiency ranges between 0.75-0.82 and 0.72-0.81 during calibration and validation periods respectively. The root mean square error ranges between 23-38 mm/month and 25-44 mm/month during calibration and validation periods respectively.

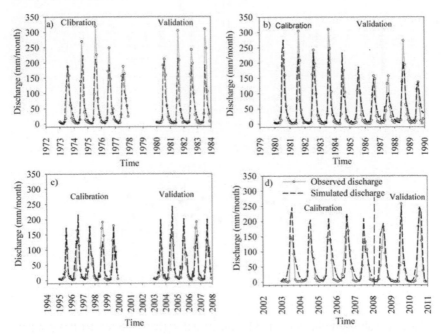

Figure 5.8: Observed and simulated monthly streamflow of Jedeb catchment for four different periods.

These results in general indicated that the model is satisfactorily reproducing the observed streamflow. Dotty plots showing Nash-Sutcliffe efficiency as a likelihood measures for parameter uncertainty in different calibration periods and the posterior cumulative distribution of 100 best performing optimal parameter sets are shown in figure 5.9 and figure 5.10, respectively.

The change in parameters over different periods is visible from the scatter dotty plots (figure 5.9). Moreover, it is seen that except the ground water recession constant, the discharge predictions are sensitive to the three parameters ($S_{max}$, $\alpha_1$ and $\alpha_2$). It is likely that the under estimation of the discharge peak is related to the poor spatial coverage of precipitation data within the catchment, though parameter uncertainty or model structural errors could be another reasons.

The dotty plots in figure 5.9 indicate the uncertainty of model parameters. It is shown in the figure that parameters are well identifiable for $S_{max}$ (storage), $\alpha_1$ (direct runoff), and $\alpha_2$ (evaporation) during the period 1973-1977. However, the recession parameter (d) shows less identifiability. The model structural error could be the reason for unidentifiability (Tekleab et al., 2011).

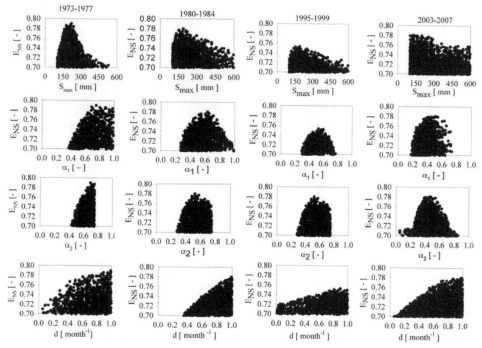

Figure 5.9: Dotty plots indicating parameter uncertainty for different calibration periods.

From the cumulative distribution, it can be seen that the distribution values of the $S_{max}$ [mm] parameter (figure 5.10a) in different time period exhibit reduction over different calibration periods. The soil moisture capacity is related with the soil infiltration rate, land cover, and topography characteristics in the catchment. The reduction of this parameter over time might be related with less infiltration due to intensive agricultural farming practices in the steep slope topography and enhance surface runoff generation. The result in turn supported by the results of flow variability analysis in terms of the change in hydrograph characteristics over time (see section 5.5.2 and 5.5.3). The catchment rainfall retention parameter $\alpha_1$ [-] depict consistent decrease over the calibration periods.

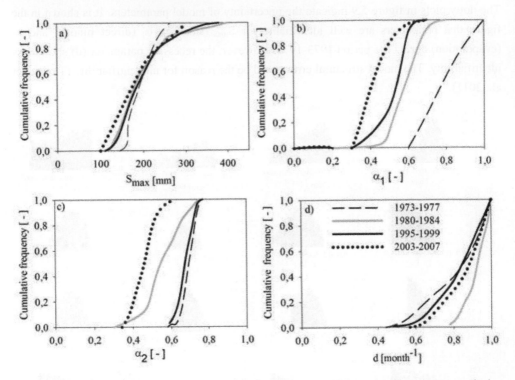

Figure 5.10: Cumulative posterior distribution of best 100 optimal parameters during different calibration periods

For example, the frequency of the peak parameter values in the 1970s were between 0.7 and 0.8 and were reduced to 0.3 and 0.4 in the 2000s which clearly exhibit the ability of the soil to retain the given rainfall amount is decreasing over time. This implies that more surface runoff generation is apparent over subsequent periods and the ground water recharge is decreasing. The physical interpretation of $\alpha_1$ [-] is related with the soil infiltration rate in the catchment. Higher value of this parameter indicates higher rainfall retention by the soil in the catchment. While the lower values of this parameter show lower rainfall retention by the soil in the catchment (Zhang et al., 2008).

For the evaporation parameter $\alpha_2$ [-] the peak parameter values in 1970s and 1990s are almost in a similar ranges between 0.64-0.75 suggesting higher partitioning of available water for evaporation. Higher values of $\alpha_2$ indicate relatively high forest cover and smaller values represent short grass cover and crops (Zhang et al., 2001a, 2004, 2008). However, it is not possible to interpret this parameter in relation with the vegetation characteristics using this simple model. The decrease in the frequency of parameter peak values in the later period of the 2000s reflects a partitioning of the available water in the soil for evaporation demand was decreasing. This may further suggest an increase in surface runoff that has led to a reduced infiltration rate of the soil and it seems likely that the evaporation is constrained by the available water in recent periods due to the restricted soil moisture in the root zone.

However, accurate estimation of the evaporation flux based on ground measurements or remote sensing approaches e.g. Bastiaanssen et al. (1998) and Su, (2002), combined with distributed hydrological modelling is essential to account explicitly the effects of evaporation from different land covers on streamflow.

### 5.5.4 Evaluation of the flow regime based on the flow duration curve

The results from the monthly flow duration curve analysis for different time periods elucidate that the high flows, represented by $Q_5$ (the flow equal or exceeded 5% of the time) in the Jedeb catchment, decreased by 3% between 1970s and 1980s, 28% decrease between 1980s and 1990s and about 45% increase between 1990s and 2000s. The majority of the median flow particularly, in 1980s was declining due to the severe drought all over the country (Conway. 2000). Similar conditions of decreasing flow in the later period also support significant decreasing trends of pulses for 15M, 20M and 25M thresholds (see section 5.5.2).

The low flow condition in different time periods exhibit consistent declining in magnitude of flows for subsequent periods. The results from investigation of 95% of flow exceedance probability reveal a general decrease for successive time periods. Magnitudes of $Q_{95}$ (flow equal or exceeded 95% of the time) are decreasing with 15% from 1970s to 1980s, 39% from 1980s to 1990s and 71% in the later period 1990s to 2000s. The reduction in $Q_{95}$ is in agreement with study by Rientjes et al. (2011b) in the Gilgel Abay catchment, Ethiopia. The strong decrease in $Q_{95}$ in the later period is to a certain extent ascribed to water abstraction for irrigation development in the Jedeb catchment since the late 1990s onwards (Tekleab et al., 2013). These results from the flow duration curve are corroborating the results of the statistical approach in terms of the changes in flow regime over the considered period.

### 5.5.5 Effects of land use/land cover (LULC) change on streamflow

Referring to table 5.2, the agricultural area expansion was greater between 1957 and 1972 than between 1972 and 2009. From the land use pattern it is noted that for the last 50 years, the conversion of different land use type into cultivated land is only 17%. Even the increment in cultivated land expansion was stopped, since 1994 suggests suitable cultivated land for further expansion was limited (Teferi et al., 2013). The temporal patterns of land use classification results depict that changes particularly forest into cultivated land had been taken place already in 1950s. Moreover, it is seen that insignificant percentage increment in cultivated land, since 1970s until 2000s and introduction of new Eucalyptus globules plantation expanded between the year 1986-1994 in smaller extent. Thus, owing to the dominance of agricultural farming practices, the streamflow regimes in Jedeb catchment might have changed.

The two different approaches used elucidate common evidence that LULC change in Jedeb has impacted streamflow characteristics. The first approach on flow variability analysis shows a significant change of flow rapidity (flashiness), frequency, duration, rise and fall rate, and an increase in number of discharge pulses in the later period (2000s). The second approach depict that there is a general decrease in the catchment rainfall retention by the soil over time.

The results obtained in the first approach (flow variability), were supported by model results that the rainfall retention capacity of the soil in Jedeb catchment decreased during last four decades, (section 5.5.1 to 5.5.3). The study re-confirmed the same results by calibrating model separately for the different time periods (section 5.5.3). The comparison of model parameters, shows enhanced peak discharge, decreased available soil moisture and decreased infiltration rate during the last four decades. This is typically associated with LULC changes and farming practices in the catchment. Similar results were shown in the literature for other regions (e.g. Sullivan et al., 2004; Moore and Heilman, 2011).

The results of this study are in agreement with Bewket and Sterk, (2005) and Rientjes et al. (2011b) with regard to the significant decrease in low flows and the increase in extreme high flows, which was attributed to land use change in the region. However, it is in contrast with the later study that the streamflow change in this study is not due to precipitation change. As mentioned in section 5.5.2 precipitation did not show statistically significant increasing or decreasing trends at annual, monthly or seasonal time scales in the Blue Nile basin including the climate station represent Jedeb catchment.

Overall, from a hydrological point of view, changes in streamflow are noticeable in the Jedeb catchment. Agricultural land use management, particularly farming practices in unfavourable slope areas, ploughing along the slope direction and less adoption of soil conservation measures e.g. Temesgen et al. (2012) preclude infiltration processes due to plough pan effect. Consequently, more surface runoff generation, lower base flow, and enhanced peak flows are the consequences due to land management practices in the Jedeb catchment. Since the prominent land use changes related to forest clearing had been taken place before 1950s, the intensification of agriculture is likely attributed to the aforementioned hydrological changes in the catchment. Therefore, from our evidence of no discernible significant trends in precipitation, and the results of combined statistical and modelling approaches, the dynamics of changes in the flow hydrograph is likely linked to the agricultural land use management practices. Hence, land use change due to farming practices have more pronounced role than climatic variability i.e. (rainfall) in the Jedeb catchment.

The approaches in this study are not without limitations. The key limitation is that the spatial and temporal variations of vegetation patterns were not accounted for by the lumped model. The LULC changes were inferred from change detection in model parameters over time.

Thus the lumped model lacks to explicitly account the spatial and temporal variations of vegetation characteristics which in turn affect the actual evaporation in the catchment. However, our results are reasonable in a sense that visible changes have seen in the streamflow due to LULC changes related with accelerated expansion of agriculture, and farming practices.

## 5.6 Conclusions

The purpose of this investigation was to quantify the hydrologic response of the Jedeb meso-scale catchment to land use/land cover changes over the period 1973-2010. The methodological approaches described in this chapter enable to comprehend how the agricultural land use management practices alter the daily flow hydrograph characteristics, the extreme flows and the reduction in soil moisture retention capacity. The results indicated that peak discharge pulses are more visible in later period, and the flow hydrograph depict increasing flashiness, flow reversal, rise and fall rate. Furthermore, the duration of discharge pulses and rainfall retention capacity of the soil are decreasing over the last four decades. These incidents are found to be the major hydrological changes observed in the Jedeb catchment.

The high flow is decreasing in the 1970s and 1980s and showed a strong increase of about 45% between the 1990s and 2000s. The reduction in low flow is much larger, 15% between the 1970s and 1980s, 39% from the 1980s to 1990s and 71% between the 1990s and 2000s. These are considerable changes in the flow regime, and may have undesirable connotation in terms of moisture stress in the irrigated crops, shortage of water for livestock and household consumptions, in which the society in the catchment largely depends on the base flow in the river.

Efforts were made to investigate the impacts of land use change on the hydrologic response of the Jedeb catchment using statistical flow variability analysis and monthly conceptual hydrological modelling approaches. However, there is a need for further research to quantify land use change impacts on streamflow using a semi distributed process based model, which accounts for the spatio-temporal variation of climatic and vegetation patterns in the catchment. Furthermore, accurate estimation of actual evaporation from ground measurements and using remotely sensed data based on an energy balance approach would enable to assess the potential impacts of vegetation change on streamflow.

In general from the perspective of sustainable water management, adverse effects should be mitigated through intervention of catchment management. Thus, the catchment management aimed at implementing soil conservation measures to reduce surface runoff and augmenting groundwater recharge in the Jedeb catchment is inevitably essential.

# CHAPTER 6

Characterisation of stable isotopes to identify residence times and runoff components in two meso-scale catchments in the Abay/Upper Blue Nile basin, Ethiopia[1]

Previous chapters presented, the hydro-climatic variables, catchment water balance, and the hydrologic responses to land use change in a case study catchment. To gain further insights into the runoff generation of the headwater catchments, the benefit of stable environmental isotopes, are explored in this chapter. Thus, to improve knowledge on runoff generation processes, stable environmental isotopes samples of precipitation, spring waters, and streamflows were collected and analysed from Chemoga and Jedeb meso-scale headwater catchments. The analysed data sets are used to characterize the stable isotopes over different altitudinal gradients within the Choke mountains range. Furthermore, a classical two component mixing model and sine wave regression approaches have been implemented to separate the different runoff components and to estimate the mean residence times in the Chemoga and Jedeb catchments. Results from the hydrograph separation on a seasonal timescale demonstrated the dominance of event water with an average of 71% and 64% of the total runoff during the wet season in the Chemoga and Jedeb catchment, respectively. The results further showed that the stable isotope compositions of streamflow samples were damped compared to the input function of precipitation for both catchments. This damping was used to estimate mean residence times of stream water of 4.1 and 6.0 months at the Chemoga and Jedeb catchment outlet, respectively. Short mean residence times and high fractions of event water components recommend catchment management measures aiming at the reduction of overland flow/soil erosion and increasing of soil water retention and recharge to enable sustainable development in these agriculturally dominated catchments.

---

[1] *This chapter is based on*: Tekleab, S., Wenninger, J., and Uhlenbrook, S. Characterisation of stable isotopes to identify residence times and runoff components in two meso-scale catchments in the Abay/Upper Blue Nile basin, Ethiopia. (2014). Hydrol. Earth Syst. Sci., 18, 2415–2431, doi: 10.5194/hess-18-2415.

## 6.1 Introduction

Environmental isotopes as tracers are commonly applied for examination of runoff generation mechanisms on different spatial and temporal scales (e.g. Uhlenbrook et al., 2002; Laudon et al., 2007; Didszun and Uhlenbrook, 2008). Isotope tracer studies are used for hydrograph separations (Sklash and Farvolden, 1979; Buttle, 1994), provide additional information for identifying source areas, and flow pathways under different flow conditions, and estimate the mean residence time of a catchment (Soulsby et al., 2000; Uhlenbrook et al., 2002; McGuire et al., 2005; McGuire and McDonnell, 2006; Soulsby and Tetzlaff, 2008). The application of isotopes in catchment hydrology studies has been carried out in small experimental catchments to meso-scale catchments (e.g. McDonnell et al., 1991; Uhlenbrook et al., 2002; Tetzlaff et al., 2007a) and large-scale catchments (Taylor et al., 1989; Liu et al., 2008a).

Only few studies have been undertaken to characterise water cycle components using stable isotopes in Ethiopia (e.g. Rozanski et al., 1996; Kebede, 2004; Levin et al., 2009; Kebede and Travi, 2012). The results from these studies indicate that stable isotope composition of precipitation is only little affected by the typical dominant controls: amount, altitude and continental effects. Despite low mean annual temperature and high altitude, the Ethiopian meteoric water (e.g. the Addis Ababa station of the International Atomic Energy Agency (IAEA)) exhibits less negative isotopic composition as compared to the East African meteoric water (e.g. Nairobi, and Dar es Salaam) (Levin et al., 2009; Kebede and Travi, 2012). The literature suggests different reasons for the less negative isotopic composition of the Addis Ababa station in Ethiopia. For instance, Joseph et al. (1992) indicated that the moisture from the Indian Ocean that results in an initial stage of condensation vapour, which did not undergo a major rainout fractionation effect, is likely the main reason for higher isotopic composition. Levin et al. (2009) hypothesised that the less negative isotopic composition in Addis Ababa is due to advection of recycled moisture from the Congo basin and the Sudd wetland. Rozanski et al. (1996) and Darling and Gizaw (2002) show that the increased sea surface temperature at the moisture source, and evaporation condition at the sources, are attributed to the less negative isotopic composition for the Ethiopian meteoric water as compared to the East African meteoric water.

Despite the paramount importance of water resources in the Abay/Upper Blue Nile area for the whole Nile basin, the usefulness of stable isotope data for catchment hydrological studies is largely unexplored. Very little is known about the use of stable isotopes for hydrological studies in the region. However, stable isotopes have the potential to provide enormous benefit with respect to hydrological process understanding and sustainable planning of water resource management strategies and policies in such data-scarce areas (Hrachowitz et al., 2011a). This potential has been demonstrated in numerous other regions worldwide (Kendall and Caldwell, 1998; Kendall and Coplen, 2001; Gibson et al., 2005; Barthold et al., 2010; Kirchner et al., 2010).

74

The use of isotope tracer techniques to understand mean residence times (MRTs) and a residence time distribution (RTDs) has received a lot of attention (Rodgers et al., 2005a; McGuire and McDonnell, 2006). They are used to gain a better understanding of flow path heterogeneities (Dunn et al., 2007), to get insights into the internal processes of hydrological systems, and are used as a tool for hydrological model construction and evaluation (Uhlenbrook and Leibundgut, 2002; Wissmeier and Uhlenbrook, 2007, Hrachowitz et al., 2011b). Furthermore, they can be used as fundamental catchment descriptors, providing information about the storage, flow pathways and sources of water (McGuire and McDonnell, 2006), and are used for conceptualising the differences in hydrological processes by comparing different catchments (McGuire et al., 2005; Soulsby et al., 2006; Tetzlaff et al., 2009). The steady state assumption of fluxes for estimating the mean residence time has been commonly used, although time-invariant mean residence times do not exist naturally in real-world catchments. To circumvent this issue a few recent studies developed a method for estimating time-variable mean residence times (e.g. van der Velde et al., 2010; Botter et al., 2011; Heidbüchel et al., 2012; Hrachowitz et al., 2013a).

In this study, the investigation of meteoric water in the source of the Abay/Upper Blue Nile basin is undertaken as the basis for characterisation and better understanding of the dominant runoff components. This is used as a baseline study for future hydrological studies using environmental isotopes in the region. The main objectives of this study are: a) to characterise the spatio-temporal variations in the isotopic composition in precipitation, spring and stream water; b) to estimate the mean residence time of stream water; and c) to separate the hydrograph on a seasonal timescale in the two adjacent meso-scale catchments, Chemoga and Jedeb in the Abay/Upper Blue Nile basin.

## 6.2. Study area

The Chemoga and Jedeb rivers are tributaries of the Abay/Upper Blue Nile basin, located south of Lake Tana, and extend between approximately 10° 10' and 10° 40'N latitude and 37° 30' and 37° 54'E longitude. Both rivers originate from the Choke Mountains at an elevation of 4000 m. a.s.l. (see Figure 6.1). The climate in these catchments has distinct seasonality with three seasons: (i) Summer as the main rainy season from June to September, (ii) winter as the dry season from October to February, and (iii) spring as the short rainy season from March to May (NMSA, 1996). The long-term average annual temperature over the period of 1973-2008 at Debre Markos weather station is about 16.3°C. The precipitation ranges between 1342 and 1434 mm a$^{-1}$ (1973-2010) in the lower and upper parts of the catchments.

In these two catchments subsistence farming is commonly practiced. The farmers rely on rain-fed agriculture for their livelihoods. Barley, oats and potato are the main crops grown in the upland area, whereas wheat, tef and maize are grown in the middle and lower parts of the catchments. According to the studies by Bewket and Sterk (2005) and Teferi et al. (2010,

2013), the land use in the Chemoga catchment had been subjected to changes before the 1950s. The major change was the increase in cultivated area at the expense of open grazing area and a slight increase in plantation forest cover due to eucalyptus plantations. Natural vegetation cover can hardly be found in both catchments. A recent study by Teferi et al. (2013) showed that 46% of the Jedeb catchment experienced transitions from one land cover to another over the last 52 years. Nowadays, about 70% of the land is used for agriculture and 3% is forest plantations and the remaining percentages are utilised for other land uses (pasture land, bare land with shrubs and bushes).

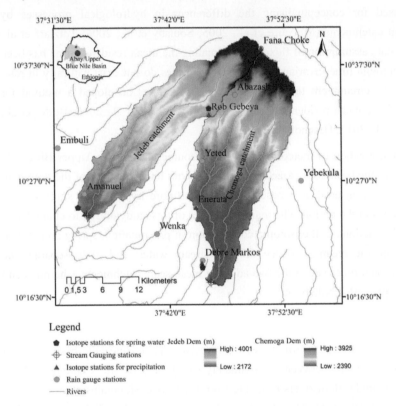

Figure 6.1: Location of the study area indicating the network of rain gauges, streamflow gauges and sampling points for stable isotopes of precipitation, surface water and spring water. The red dot within the Ethiopian map indicates the location of the Chemoga and Jedeb catchments.

## 6.3 Methodology

### 6.3.1 Hydro-meteorological data collection

Streamflow data sets are based on manual water level measurements (daily at 06:00 a.m. and 06:00 p.m.) at the Chemoga and Jedeb gauging stations from 2009-07-01 to 2011-08-31.

Based on the stage discharge relationships rating curves were developed using regression models (see detail in chapter 7).

A network of thirty two manual rain gauges was established, since July 2009. Unfortunately, only ten rain gauges are working properly (see sample of manual rain gauge fig. 6.2). From these ten rain gauges, daily precipitation data were collected over the same period as the stream flows and daily temperature data at Debere Markos station over the same period were obtained from the Ethiopian National Meteorological Agency. The temperature data at Debre Markos station was used to estimate the temperature at Enerata, Rob Gebeya, Fana Choke, and Yewla stations based on a decrease of 0.6°C in temperature per 100 m increase in altitude. The catchment average precipitation amount, catchment mean annual temperature, potential evaporation, and isotopic composition of precipitation were computed using the Thiessen polygon method. Due to the limited climatic data availability, the potential evaporation was computed using the Hagreaves method (Hargreaves and Samani, 1982). The method was selected due to the fact that other meteorological data (e.g. humidity, solar radiation, wind speed, etc.) were missing and only temperature data were available for the study catchments. The intra-annual variability of hydro-climatic data within the catchments is shown in Figure 6.3. Furthermore, detailed description of the hydro-meteorological data is presented in Table 6.1.

Table 6.1: Descriptions of hydro-meteorological characteristics of investigated catchments (2008-2010). P, Q, $E_p$, E stand for catchment average precipitation, runoff, potential evaporation and actual evaporation, respectively.

| Catchment | Area (km$^2$) | Mean annual values (mm a$^{-1}$) | | | | | Mean annual temperature (°C) |
|---|---|---|---|---|---|---|---|
| | | P | Q | $E_p$ | E | T | |
| Chemoga | 358 | 1303 | 588 | 1338 | 715 | 13.9 | |
| Jedeb | 296 | 1306 | 692 | 1384 | 614 | 15 | |

Figure 6.2: Sample manual rain gauge (left) and Precipitation water samples collector for isotope analysis (right). The inset figure right is the 2 mL sampling bottles for the different water samples (precipitation, springs, and river).

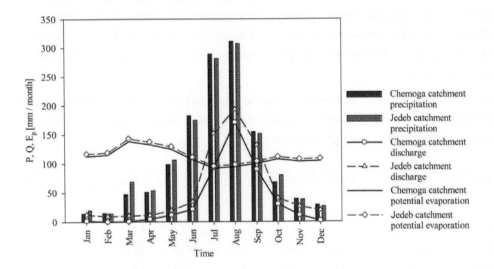

Figure 6.3: Intra-annual variability of hydro-climate data for the period 2008-2010 showing similar climate and distinct streamflow response in the Chemoga and Jedeb catchments. P, Q and $E_p$ on the y-axis stand for precipitation, discharge and potential evaporation, respectively.

## 6.3.2 Field measurements and sampling

To characterise the spatial and temporal variability of stable isotope composition in precipitation, spring discharge and streamflow, field investigations were undertaken, from August 2008 until August 2011. At five different locations, plastic funnels were used to collect the precipitation samples for the analysis of isotopic signature in precipitation. The samples were collected on a bi-weekly basis.

The rainfall sample collectors have a capacity of 10 litres fitted with a vertical funnel with a mesh on top to avoid dirt and a long plastic tube to minimise evaporation out of the collection device according to the IAEA (2009) technical procedure for sampling (see fig. 6.2). Spring water was sampled at three locations at different altitudes on a weekly basis, and the two weekly samples were mixed and taken for the analysis. Streamflow was sampled at the outlet of the Chemoga and Jedeb rivers on a weekly basis. During sampling, the water was filled into 2 mL glass bottles and closed immediately to avoid fractionation due to evaporation. Details about isotope sample locations and investigation periods are given in Table 6.2.

Table 6.2: Description of isotope sample locations and total number of samples taken during the investigation period (Aug.2008-Aug.2011).

| Sample type | Location name | Abbreviation | Elevation (m. a.s.l) | Number of samples | Investigation period |
|---|---|---|---|---|---|
| Precipitation | Debre Markos | P-DM | 2515 | 58 | Aug. 2008- Aug 2011 |
| Precipitation | Enerata | P-EN | 2517 | 24 | July 2009- Aug. 2011 |
| Precipitation | Rob Gebeya | P-RG | 2962 | 53 | Oct. 2008-Aug. 2011 |
| Precipitation | Fana Choke | P-FC | 3993 | 41 | July 2009-Aug. 2011 |
| Precipitation | Yewla | P-YW | 2219 | 46 | July 2009-Aug. 2011 |
| Spring | Debre Markos | S-DM | 2339 | 64 | Aug.2008- Aug. 2011 |
| Spring | Rob Gebeya | S-RG | 2820 | 53 | Oct. 2008-Aug. 2011 |
| Spring | Yewla | S-YW | 2255 | 59 | July 2009-Aug 2011 |
| Stream | Chemoga | Q-CH | 2402 | 83 | July 2009-July 2011 |
| Stream | Jedeb | Q-JD | 2190 | 98 | July 2009-July 2011 |

### 6.3.3 Laboratory analysis

All water samples were analysed at UNESCO-IHE (Delft, the Netherlands) using an LGR liquid-water isotope analyser. The stable isotopic composition of oxygen-18 and deuterium are reported using the $\delta$ notation, defined according to the Vienna Standard Mean Ocean Water (VSMOW) with $\delta^{18}O$ and $\delta^2H$. The accuracy of the LGR liquid-water isotope analyser measurements was 0.2‰ for $\delta^{18}O$ and 0.6‰ for $\delta^2H$, respectively.

The accuracy of the LGR liquid-water isotope analyser measurements was 0.2‰ for $\delta^{18}O$ and 0.6‰ for $\delta^2H$, respectively.

### 6.3.4 Hydrograph separation on a seasonal timescale

The classical steady state mass balance equations of water and tracer fluxes in a catchment were used in this study to separate the hydrograph into different components. The assumptions used for the hydrograph separations and the basic concepts are described in detail by e.g. Sklash and Farvolden (1979), Wels et al. (1991) and Buttle (1994).

The mass balance equation used for a time-based two-component separation using ($^{18}$O) as a tracer can be described as

$$Q_T = Q_E + Q_{Pe} \tag{6.1}$$

$$C_T Q_T = C_E Q_E + C_{Pe} Q_{Pe}, \tag{6.2}$$

Where $Q_T$ is the total runoff [m³s⁻¹], and $Q_E$ [m³s⁻¹] and $Q_{Pe}$ [m³s⁻¹] are the runoff event and pre-event components, respectively. $C_T$ is the total concentration of tracer observed in total runoff [‰ VSMOW], and $C_E$ [‰ VSMOW] and $C_{Pe}$ [‰ VSMOW] are the tracer concentrations in event and pre-event water, respectively. Combining equation 6.1 and equation 6.2, the contribution of event water and pre-event water to the total runoff can be estimated as:

$$Q_E = Q_T \left( \frac{C_T - C_{Pe}}{C_E - C_{Pe}} \right) \tag{6.3}$$

$$Q_{Pe} = Q_T \left( \frac{C_T - C_E}{C_{Pe} - C_E} \right) \tag{6.4}$$

The precipitation isotopic composition was weighted based on the cumulative incremental weighting approach as outlined by McDonnell et al. (1990):

$$\delta^{18}O = \frac{\sum_{i=1}^{n} p_i \delta_i}{\sum_{i=1}^{n} p_i} \tag{6.5}$$

Where $p_i$ and $\delta_i$ denote the rainfall amount and $\delta$ value, respectively.

Similarly, the monthly discharge isotopic composition in the rivers was weighted using equation 6.6.

$$\delta^{18}O = \frac{\sum_{i=1}^{n} Q_i \delta_i}{\sum_{i=1}^{n} Q_i} \tag{6.6}$$

Where, $Q_i$ [m³ s⁻¹] is the daily volumetric flow rate and $\delta_i$ [‰] is the isotopic composition of the streamflow.

Due to the distinct seasonality, the precipitation during the dry (winter) and little rain (spring) seasons does not contribute significantly to the total streamflow, neither as surface nor as subsurface flow. This is due to the fact that the precipitation in these seasons mostly evaporates without producing direct runoff or recharging the groundwater (Kebede and Travi, 2012). To account for the effects of seasonality on the results of hydrograph separation, the end member signature is not taken as a constant value throughout the whole seasons. Consequently, the pre-event water isotopic composition was taken as the monthly isotopic values at each month during the dry (winter) and spring seasons. At the same time, to see the effects of different pre-event end member concentrations on the results of hydrograph separation, three different end members were estimated: First, the average values for the whole dry season isotope concentration were taken as all end members, second, the average value of the isotopic concentration in the month of February, which represents the baseflow in the rivers, was considered as an end member, and third, the average isotopic concentration of combined dry and spring season concentrations.

The event water $\delta^{18}O$ end member was taken as the weighted mean isotopic composition of precipitation, in each month for the investigated period. The differences in isotopic composition for event water vary from -6.37 to -4.24‰, and pre-event water from -0.25 to 0.62‰ is adequate for the hydrograph separation in these catchments based on the assumptions of classical hydrograph separation described in Buttle (1994).

Hydrograph separation using isotope technique is prone to error due to the uncertainty in the estimation of end member concentrations (e.g. Genereux, 1994; Uhlenbrook and Hoeg, 2003). In this study the uncertainty in the two-component separations during the wet season June to September is evaluated based on the Gaussian error propagation technique according to equation 6.7 (e.g. Genereux, 1994).

$$W_y = \sqrt{(\frac{\partial y}{\partial x1}W_{x1})^2 + (\frac{\partial y}{\partial x2}W_{x2})^2 + ...... + (\frac{\partial y}{\partial xn}W_{xn})^2} \qquad (6.7)$$

Where W represents the uncertainty in the variables indicated in the subscript, assuming that 'y' is a function of the variables $x_1$, $x_2$...,$x_n$ and the uncertainty in each variable is independent of the uncertainty in the others (Genereux, 1994). The uncertainty in y is related to the uncertainty in each of the subscript variables by using equation 6.7. Application of equation 6.7 into equation 6.4 gives the propagated total uncertainty related to the different component computed using equation 6.8.

$$W = \left\{ \left[ \frac{(C_E - C_T)}{(C_E - C_{Pe})^2} * W_{CPe} \right]^2 + \left[ \frac{(C_T - C_{Pe})}{(C_E - C_{Pe})^2} * W_{CE} \right]^2 + \left[ \frac{-1}{(C_E - C_{Pe})} * W_{CT} \right]^2 \right\}^{\frac{1}{2}} \qquad (6.8)$$

Where W is the total uncertainty or error fraction related to each component, and $W_{CPe}$, $W_{CE}$, and $W_{CT}$ are the uncertainty in the pre-event, event and total stream water, respectively. The uncertainties related to each component are computed by multiplying the standard deviations by $t$ values from the Student's $t$ distribution at the confidence level of 70% (Genereux, 1994).

### 6.3.5 Estimation of mean residence time

The mean residence time of stream water in a catchment is commonly computed using lumped parameter black box models described in Maloszewski and Zuber (1982). However, the application of this method to short data records and coarse spatial and temporal sampling leads to inaccurate estimates of parameters and tracer mass imbalance, if the timescale of residence time distribution is larger than the input data (McGuire and McDonnell, 2006).

Hence, due to the short record length and coarse frequency of spatial and temporal tracer sampling, in this study the mean residence time is estimated based on the sine wave approach fitting the seasonal $\delta^{18}O$ variation in precipitation and streamflow (e.g. McGuire et al., 2002; Rodgers et al., 2005a; Tetzlaff et al., 2007b). The method gives indicative first approximation estimates of mean residence times (Soulsby et al., 2000; Rodgers et al., 2005a). The predicted $\delta^{18}O$ can be defined as

$$\delta = C_o + A\left[\cos\left(C_f t - \Phi\right)\right] \tag{6.9}$$

Where $\delta$ is the predicted $\delta^{18}O$ [‰] composition, $C_o$ is the weighted mean annual measured $\delta^{18}O$ [‰], A is the annual amplitude of predicted $\delta^{18}O$ [‰], $C_f$ is the angular frequency constant ($0.017214$ rad $d^{-1}$), t is the time in days after the start of the sampling period and $\Phi$ is the phase lag of predicted $\delta^{18}O$ in radians. Furthermore, equation 6.9 can be evaluated using sine and cosine terms in a periodic regression analysis (Bliss, 1970) as:

$$\delta = C_o + \beta_{\cos} \cos(C_f t) + \beta_{\sin} \sin(C_f t) \tag{6.10}$$

The estimated regression coefficients $\beta_{\cos}$ and $\beta_{\sin}$ are used to compute the amplitude in input, and output signals $\left(A = \sqrt{\beta^2 \cos + \beta^2 \sin}\right)$ and consequently the phase lag $\tan \Phi = \left|\dfrac{\beta_{\sin}}{\beta_{\cos}}\right|$.

The mean residence time from the fitted sine wave in input and output signals was estimated as

$$T_{rs} = C_f^{-1}\left[\left(\frac{A_2}{A_1}\right)^{-2} - 1\right]^{0.5} \tag{6.11}$$

Where $T_{rs}$ is the mean residence time [d], $A_1$ is the amplitude of precipitation $\delta^{18}O$ [‰], $A_2$ is the amplitude of streamflow $\delta^{18}O$ [‰], and $C_f$ is defined in equation 6.9.

## 6.4 Results and discussion

### 6.4.1 Meteoric water lines

The plot representing the relationship between $\delta^{18}O$ and $\delta^2H$ isotopic composition for precipitation is shown in Fig.6.4. The spring and river waters isotopic compositions are also plotted in the same figure for comparison. The spatial distribution of $\delta^{18}O$ and $\delta^2H$ composition of precipitation varies considerably along the elevation gradient. Fana Choke station located at the highest elevation has more negative isotope compositions than the Yewla lower altitude station. The air masses lifted at the higher altitude with lower air temperature and higher humidity could be possible reasons for elevation-dependent variations of isotope composition. The difference in the isotopic composition at these two stations was evaluated using the Wilcoxon signed rank statistical test. The test results show that the difference in the isotopic values at the two locations is statistically significant ($p = 0.020$) evaluated at the 95% confidence level. The scatter of the isotopic composition from the global meteoric water line might be related to the effect of the evaporation of falling rain drops, condensation in the cloud and different moisture sources over different seasons (Dansgard, 1964; Gat, 1996; Levin et al., 2009; Kebede and Travi, 2012). However, from the plot of the relationship between $\delta^{18}O$ and $\delta^2H$ isotopic composition for precipitation, the effect of evaporation is insignificant and plots along the local evaporation line. This means that the isotope values do not deviate from the local and global water lines.

The representation of the isotope values with the local meteoric water line was compared with that of the Addis Ababa Meteoric Water Line (MWL) produced using Global Network of Isotopes in Precipitation (GNIP) samples (Fig. 6.4). The relationships of $\delta^{18}O$ and $\delta^2H$ composition of the present study all exhibit an almost similar slope to that of the Addis Ababa station. However, it has a higher intercept/deuterium excess than the Addis Ababa GNIP station. The higher d excess values > 10‰ in both Addis Ababa and the study area are attributed to land surface–atmosphere interaction through transpired moisture contribution (Gat et al., 1994).

Furthermore, it is shown that the precipitation waters with more positive isotopic values are derived from the winter and spring season precipitation, whereas those plotting at the more negative end of the LMWL are derived from summer precipitation. The river water isotopic values in the Chemoga and Jedeb catchments exhibit little variation along the LMWL. These variations indicate that the waters for both catchments derived mostly from the summer precipitation. The spring waters at Debre Markos and Rob Gebeya exhibit negative isotopic composition as compared to Yewla, which is located towards the positive end of the LMWL.

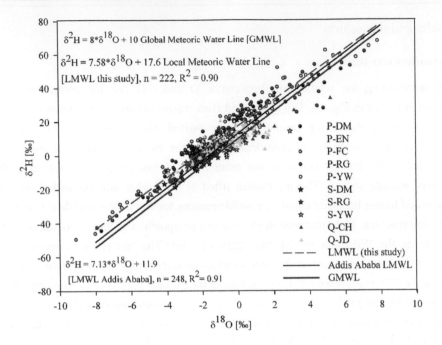

Figure 6.4: Relationship between, $\delta^{18}O$ and $\delta^2H$ for precipitation, stream and spring water in the study area. The abbreviations in the legend are described in Table 6.2.

### 6.4.2 Spatio-temporal variation of isotope composition in precipitation, spring water and streamflow

**Isotope composition of precipitation**

The results of the measured isotopic composition of precipitation samples exhibit marked spatial and seasonal variations (Fig. 6.5). The precipitation at Yewla station (lowest altitude) shows less negative $\delta^{18}O$ and $\delta^2H$ values in contrast to more negative isotopic composition at Fana Choke station (highest altitude). This shows the anticipated isotopic composition influenced by the altitude effect (Dansgaard, 1964; Rozanski et al., 1993). Nevertheless, the altitude effect varies temporally over different seasons depending on the moisture source, amount and trajectories of air mass bringing precipitation and local meteorological settings (Aravena et al., 1999). For instance, the seasonal isotopic composition relationship with elevation along the gradient during different seasons shows more negative values of isotopic composition at higher altitudes (Fig. 6.6). The altitude effect accounts for -0.12‰ and -0.58‰ per 100 m increase in altitude for $\delta^{18}O$ and $\delta^2H$, respectively. This is consistent with an earlier finding by Kebede and Travi (2012), who found that more negative values of $\delta^{18}O$ by -0.1‰ per 100 m in the higher elevations of the Upper Blue Nile plateau. The $\delta^{18}O$ and $\delta^2H$

composition is also affected by the precipitation amount effect (Dansgaard, 1964; Rozanski et al., 1993).

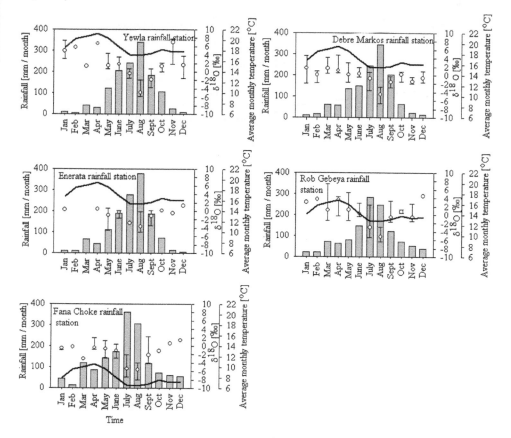

Figure 6.5: Spatial and intra-annual variations of average monthly precipitation, temperature and isotopic composition of $\delta^{18}O$ in precipitation. The error bar of the isotopic measurements stand for the standard deviation. Missing error bars for some months are due to limited isotope samples. The grey bar, black solid line and open circle with error bar are the precipitation, temperature and isotopic composition, respectively. For the isotopic composition, the open circle and the lower and upper error bars indicate the median and the 25th and 75th percentiles for the raw (non-weighted) precipitation isotope sample data, respectively.

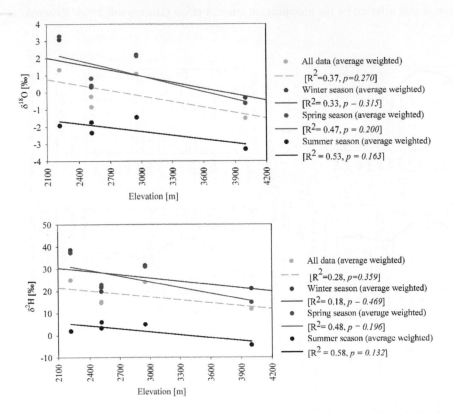

Figure 6.6: Relationships between average and seasonal amount weighted isotopic composition of precipitation with elevation at five precipitation sampling stations.

Figure 6.7 and 6.8 illustrate the $\delta^{18}O$ and $\delta^2H$ composition of precipitation at sampling stations, which show moderate regression coefficients ranging from ($R^2$=0.36-0.68, $p$ value varies from 0.001 to 0.007) for precipitation and ($R^2$ = 0.26-0.39, $p$ value varies from 0.001 to 0.007) for temperature. This suggests that the amount effect at each sampling location is important for the variation in isotopic composition in the area in addition to other factors. However, the results from this study are in contrast to the earlier studies by Kebede (2004) and Kebede and Travi (2012), who reported weak relationships between rainfall amounts and isotopic compositions in the north-western Ethiopian plateau.

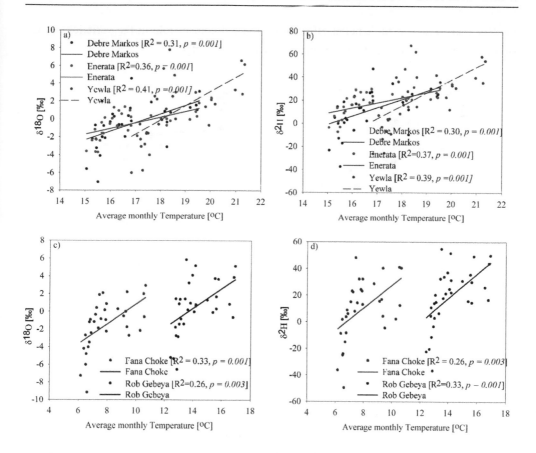

Figure 6.7: Relationships between amount-weighted isotopic compositions of precipitation samples at different stations with monthly average air temperature over the investigation period.

Moreover, multiple linear regression models are used to show the effect of monthly precipitation and mean monthly temperature on $\delta^{18}O$ and $\delta^2H$ isotopic composition of precipitation in the Chemoga and Jedeb catchments, respectively. The multiple regression models for $\delta^2H$ composition in the Chemoga and Jedeb catchments are described as

- $\delta^2H = -0.096P + 2.093T + 0.736$ ($R^2 = 0.74$, n = 28, $p = 0.001$ for precipitation and $p = 0.121$ for temperature) in the Chemoga catchment, and

- $\delta^2H = -0.116P + 2.414T - 1.374$ ($R^2 = 0.76$, n = 28, $p = 0.001$ for precipitation and $p = 0.175$ for temperature evaluated at the 5% significance level) in the Jedeb catchment.

where P in the regression equation is the monthly precipitation (mm month$^{-1}$) and T is the mean monthly average temperature (°C). These results from the multiple linear regression models also show that the amount effect has a more dominant role in the variations in the isotopic composition in the study area than the temperature effect.

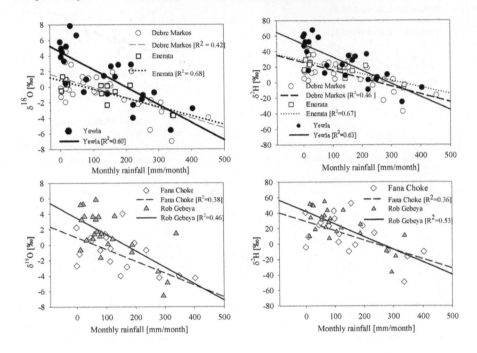

Figure 6.8: Relationship between amount-weighted isotopic compositions of precipitation samples at different stations with monthly precipitation amounts at respective stations during the investigation period.

The seasonal variations in the isotopic composition of precipitation are observed among the stations. For instance, the winter seasonal mean weighted $\delta^{18}O$ composition of precipitation has, with a value of -0.41‰, a negative isotope value at Fana Choke at the higher altitude and has positive isotope values of 3.08‰ at the lowest altitude (Yewla station). During the spring season the mean weighted $\delta^{18}O$ composition is -0.62‰ at Fana Choke and 3.3‰ at Yewla. Similarly, during summer a more negative isotopic composition of -3.28‰ is observed at Fana Choke and a relatively less negative composition -1.9‰ is observed at Yewla.

In the Chemoga catchment the mean weighted seasonal isotopic compositions of $\delta^{18}O$ and $\delta^2H$ in precipitation during winter, spring and summer are 0.72 and 24.85‰, 0.86 and 23.71‰, and -2.09 and 2.36‰, respectively. Obviously the summer seasonal isotopic compositions in both $\delta^{18}O$ and $\delta^2H$ have more negative values than in the winter and spring seasons owing to the different moisture sources and the local meteorological settings like precipitation, air temperature, and humidity. In comparison to the Chemoga catchment, the mean weighted seasonal isotopic composition in the Jedeb catchment shows consistently less negative $\delta^{18}O$ and $\delta^2H$ isotopic values of 1.48 and 29.23‰, 1.65 and 28.30‰, and -1.93 and 2.74‰ in the winter, spring and summer seasons, respectively. This implies that the less negative isotopic values of precipitation are likely related to different temperatures and altitudes in the Jedeb catchment.

**Isotope composition of spring water**

The isotopic composition of spring water at the three locations shows distinct variability ranging from -8.5 to 13.5‰ and -4.1 to 2.9‰ for $\delta^2H$ and $\delta^{18}O$, respectively (see table 6.3). The spring waters of Debre Markos (at elevation of 2339 m a.s.l) and Rob Gebeya (at 2820 m a.s.l) exhibit more negative isotopic compositions as compared to the isotopic composition of spring water of Yewla (at an elevation of 2255 m a.s.l), which showed less negative isotopic compositions. The mean raw isotopic composition of spring water indicates a wide variation at the three locations. The observed mean isotopic variation at the three locations ranged from -0.6 to 5.5‰ for $\delta^2H$ and -2.1 to -0.7‰ for $\delta^{18}O$. The mean values for $\delta^{18}O$ at Debre Markos and Rob Gebeya exhibit a similar isotopic composition of -2.1‰.

It is interesting that the isotopic compositions for the springs at Debre Markos and Rob Gebeya follow similar patterns and exhibit no major distinction in their isotope composition (Figure 6.9). This indicates that the spring water isotopic composition for both springs derived from the same altitude range of the recharge area or different areas with the same mean elevation. The mean seasonal isotopic variations at the three spring locations during the winter season ranged between -0.2 and 4.7‰, and between -2.1 and -0.7‰ for $\delta^2H$ and $\delta^{18}O$, respectively. During spring season the mean seasonal isotopic variations ranged between -2.6 and 6.3‰, and between -2.4 and -0.5‰ for $\delta^2H$ and $\delta^{18}O$, respectively. During the summer season the mean seasonal isotopic variations ranged between 0.2 and 5.9‰, and between -2.0 and -0.9‰ for $\delta^2H$ and $\delta^{18}O$, respectively. This implies that the isotopic values are more negative during the spring season and positive during the winter and summer seasons.

The mean winter and spring seasonal $\delta^{18}O$ isotopic compositions of precipitation exhibit values greater than 0‰ for all five stations (see section 6.5.2), except the more negative values at the highest altitude (Fana Choke station). In contrast to the precipitation signature during these seasons, the spring waters exhibit a more negative isotopic composition. This is suggests that the spring water during the winter and spring seasons is merely derived from summer season precipitation. The highlands seem to be the main recharge area of the spring (see figure 6.11). This shows that the winter and spring season precipitation does not contribute significantly to recharging the groundwater. This finding is in agreement with the previous studies in the region (e.g. Kebede et al., 2003; Kebede and Travi, 2012). They pointed out that during the dry and spring seasons most of the water is evaporated without contributing to the groundwater recharge.

Furthermore, the damped spring water isotopic signature as compared to the river water gives a hint that the spring water could be a mixture of old water components having longer residence times than the river water.

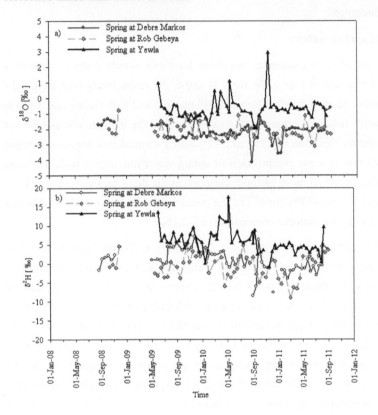

Figure 6.9: Temporal isotopic variability of three different non-weighted spring water samples: a) $\delta^{18}O$ and b) $\delta^{2}H$ composition over different investigation periods.

**Isotopic composition of river water**

The mean volume weighted $\delta^{18}O$ isotope value for the Chemoga catchment was -1.4‰, and the $\delta^{2}H$ composition was 2.7‰. During the winter or dry seasons the mean isotopes compositions were 0.1 and 6.2‰; for the $\delta^{18}O$ and $\delta^{2}H$, respectively. The mean isotope compositions during the spring or little rainy seasons were -0.7 and 11.7‰ for $\delta^{18}O$ and $\delta^{2}H$, respectively. The summer or main rainy season mean isotope compositions were -2.3 and -3.3‰, respectively. This implies that the summer isotope composition always exhibited more negative values than the winter and spring seasons.

In the Jedeb catchment the mean volume weighted $\delta^{18}O$ and $\delta^{2}H$ compositions in river water are -0.6 and 4.9‰, respectively. For the dry season (winter) the mean $\delta^{18}O$ and $\delta^{2}H$ values were 0.12 and 6.3‰, respectively. For the little rainy season (spring) the mean $\delta^{18}O$ and $\delta^{2}H$

values were -0.3 and 8.5‰, and for the summer long rainy season -1.3 and 1.7‰, respectively. These results show that in all seasons except for the $\delta^2H$ composition in the spring season, the Jedeb river water exhibits more positive isotope composition as compared to the Chemoga river. Moreover, the damped response of the isotope signature during the summer season in the Jedeb river as compared to the Chemoga river might suggest that the differences in catchment storage have relatively longer mean residence time.

Field visits during three years also supported the hypothesis that during the dry season in the Jedeb catchment the flow was sustained, while in the Chemoga catchment the dry season flow in the river was occasionally not sustained. Thus, the two catchments have different storage capacities. This in turn seems to be related to the differences in hydrologic behaviour (see fig 6.10). A water balance study by Tekleab et al. (2011) in these catchments has also shown their hydrological differences in terms of partitioning the available water on an annual timescale. The plot of the annual evaporation ratio (the ratio of mean annual evaporation to mean annual precipitation) versus the aridity index (the ratio of mean annual potential evaporation to mean annual precipitation) in a Budyko curve demonstrated a higher evaporation ratio in the Chemoga catchment than in the Jedeb catchment.

Figure 6.10 presents the temporal variations in $\delta^{18}O$ and $\delta^2H$ for the Chemoga and Jedeb catchments. In the figure the isotope composition streamflow during the main rainy season reflects damped characteristics (decreases in the amplitude of the streamflow isotope signals) as compared to the fluctuations in precipitation as were observed by e.g. McDonnell et al. (1990), Buttle (1994), and Soulsby et al. (2000). These investigations indicate that the damping behaviour of the isotope signal in streamflow is due to the fact that the pre-event or old water component of the groundwater is a mixture of many past precipitation events and resulted in an isotopic concentration which is higher than the precipitation composition during storm events. The same holds true during summer months, when rainfall generates the highest flows, in the hydrological year 2010 at both catchments, the isotope composition of river water exhibits a damped response as compared to the precipitation responses (Fig. 6.10 inset figures).

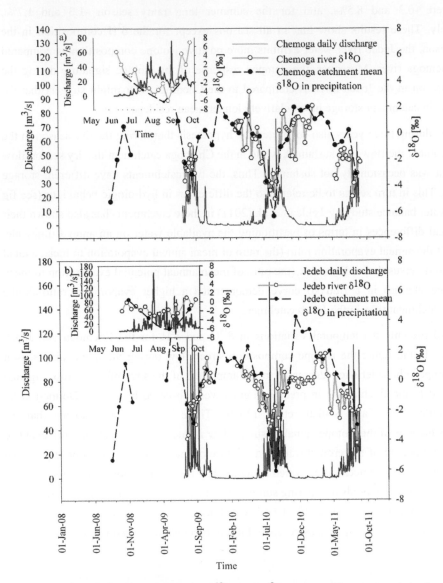

Figure 6.10: Temporal variations in $\delta^{18}O$ and $\delta^{2}H$ composition in precipitation and river discharge along with the daily flow rate for a) the Chemoga catchment and b) the Jedeb catchment. The inset figures are the details for summer season discharge and isotope composition for precipitation and streamflow for the hydrological year 2010.

### 6.4.3 Potential moisture source areas for the study area

Table 6.3 presents the mean, minimum, maximum and standard deviation of the amount-weighted precipitation and volume-weighted discharge data and the non-weighted composition for spring water. It demonstrates that more negative isotopic compositions values are observed during the main rainy season from June to September and that the less negative values are observed during the winter and spring seasons. This is obviously related to the multiple moisture sources (e.g. the Atlantic-Congo vegetation, Sudd swamp, and the Indian Ocean) and to the local meteorological processes (e.g. localised precipitation, air temperature, and humidity) (Kebede and Travi, 2012).

Table 6.3. Mean, range and standard deviation of $\delta^2H$ and $\delta^{18}O$ [‰, VSMOW] amount-weighted concentration for precipitation and volume-weighted for discharge and non-weighted for spring water during different investigation periods.

| Description | Mean [‰, VSMOW] | | Minimum [‰, VSMOW] | | Maximum [‰,VSMOW] | | Standard deviation [‰, VSMOW] | |
|---|---|---|---|---|---|---|---|---|
| | $\delta^2H$ | $\delta^{18}O$ | $\delta^2H$ | $\delta^{18}O$ | $\delta^2H$ | $\delta^{18}O$ | $\delta^2H$ | $\delta^{18}O$ |
| Precipitation at Yewla | 22.5 | 1.0 | -25.5 | -5.7 | 67.8 | 7.8 | 24.7 | 3.5 |
| Precipitation at Debre Markos | 12.3 | -0.6 | -37.1 | -7.0 | 74.6 | 4.6 | 18.4 | 2.4 |
| Precipitation at Enerata | 15.4 | -0.8 | -7.1 | -3.7 | 29.2 | 1.3 | 11.9 | 1.7 |
| Precipitation at Rob Gebeya | 23.0 | 1.0 | -36.5 | -6.5 | 54.8 | 5.9 | 21.8 | 3.1 |
| Precipitation at Fana Choke | 8.5 | -1.8 | -49.3 | -9.1 | 48.3 | 2.2 | 25.5 | 3.0 |
| Chemoga catchment precipitation | 15.5 | -0.4 | -29.8 | -6.4 | 40.2 | 3.4 | 16.7 | 2.2 |
| Jedeb catchment precipitation | 18.3 | 0.13 | -28.4 | -6.1 | 45.9 | 4.4 | 18.6 | 2.5 |
| Chemoga discharge | 2.7 | -1.4 | -15.5 | -3.9 | 19.8 | 0.8 | 9.2 | 1.5 |
| Jedeb discharge | 4.9 | -0.6 | -3.3 | -3.5 | 13.1 | 0.8 | 4.6 | 1.1 |
| Yewla spring water | 5.7 | -0.7 | -8.0 | -3.9 | 13.5 | 2.9 | 4.0 | 0.9 |
| Debre Markos spring water | 0.1 | -2.1 | -8.5 | -4.1 | 7.2 | -0.6 | 5.8 | 0.6 |
| Rob Gebeya spring water | -0.6 | -2.1 | -9.0 | -3.1 | 5.1 | -0.8 | 3.3 | 0.5 |

Seasonal variations in the isotopic composition of different water samples are shown in Fig. 11. The isotope values in different seasons might suggest different potential moisture sources bringing precipitation into the study area. These different moisture sources are investigated by mapping the potential source areas of precipitation for different seasons. Figure 6.12 presents these source areas of precipitation for different seasons, whereby the starting points of the trajectories in the study area were computed using the HYSPLIT (Hybrid Single Particle Lagrangian Integrated Trajectory) model developed by NOAA (National Oceanic and Atmospheric Administration) at the Air Resources Laboratory. The model computes the trajectories by tracing back an air package for 14 days in different seasons. (www.arl.noaa.gov/HYSPLIT_info.php).

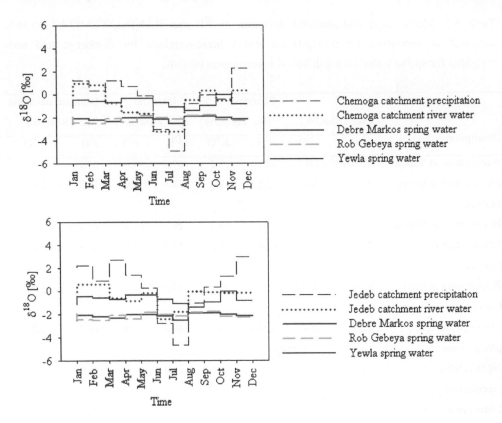

Figure 6.11: Monthly $\delta^{18}O$ [‰] isotopic variation in amount-weighted precipitation, volume-weighted discharge and non-weighted spring water a) Chemoga and b) Jedeb catchment for the period July 2009-August 2011.

It is shown that during the three seasons the source areas for starting points of moisture trajectories into the study area are different. During the main rainy season, i.e. the summer, the Atlantic Ocean, Indian Ocean, the White Nile and the Congo basin are the potential source areas of precipitation in the study area. However, in the spring and winter seasons, the potential source areas of moisture origin that are responsible for generating the little precipitation in study area is the Arabian Sea-Mediterranean sea and to some extent the Indian Ocean. These results are in agreement with the earlier studies with regard to the source areas from the Atlantic Ocean, Indian Ocean, and the Congo basin (e.g. Levin et al., 2009; Kebede and Travi, 2012).

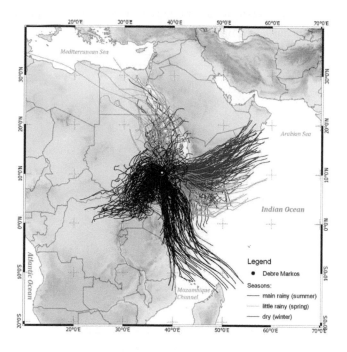

Figure 6.12: Potential source areas of precipitation to the study area in different seasons. The different lines indicate the starting points of 14-day backward calculated trajectories. The black dot indicates the location of the study area.

To date many research findings have not reached a consensus on the origin of common moisture source areas to the northern Ethiopian highlands. Mohamed et al. (2005) indicated that the moisture flux for the Northern Ethiopian plateau has mainly Atlantic origin. However, a recent moisture transport study by Viste and Sorteberg (2013) in the Ethiopian highlands reported that the moisture flow from the Gulf of Guinea, the Indian Ocean and from the Mediterranean region across the Red Sea and the Arabian Peninsula are identified as the main sources of moisture transport in the region. According to their study, the largest contribution to the moisture transport into the north Ethiopian highland was attributed to the air travelling from the Indian Ocean and from the Mediterranean region across the Red Sea and the Arabian Peninsula.

## 6.4.4 Hydrograph separation on a seasonal timescale.

The results of the two component seasonal hydrograph separations reveal that the event water fraction is more dominant than the pre-event component in both catchments, in particular during the rainy season (Fig. 6.13). The proportion of the summer (rainy season) monthly variation in the event water component varies from 32% to 99% with an average of 71% in the Chemoga catchment and 31% to 96% with an average of 64% in the Jedeb catchment over three different wet seasons of the investigation period (Table 6.4). Obviously, pre-event water is almost the sole contribution during the other seasons.

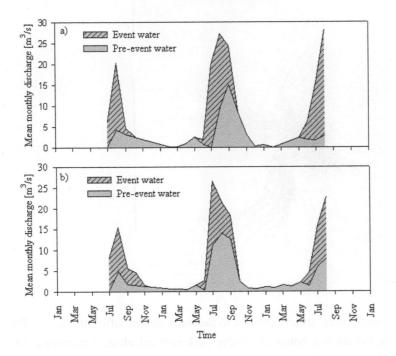

Figure 6.13: Two-component hydrograph separations using $\delta^{18}O$ as a tracer in the a) Chemoga and b) Jedeb meso-scale catchments on a seasonal timescale over the period July 2009-Aug. 2011.

The average proportions of the different runoff components due to different end members (i.e. the whole dry season average concentration and the average of the dry and little rainy season concentrations) exhibits a higher proportion of event water but vary from 62 to 67% and 33 to 38% for the pre-event water in the Chemoga catchment. In the Jedeb catchment the event water component varies from 52 to 55% and the pre-event water varies from 45 to 48%, respectively, due to different end members. These relatively small ranges of flow component contribution give a hint of the robustness of the method.

Table 6.4 Proportion of runoff components in the Chemoga and Jedeb catchments during the wet season.

| Month | Chemoga catchment | | Jedeb catchment | |
|---|---|---|---|---|
| | $Q_E$ (%) | $Q_{Pe}$ (%) | $Q_E$ (%) | $Q_{Pe}$ (%) |
| July 2009 | 94.59 | 5.41 | 96.46 | 3.54 |
| Aug. 2009 | 79.36 | 20.64 | 68.44 | 31.56 |
| Sept. 2009 | 32.03 | 67.97 | 69.34 | 30.66 |
| June 2010 | 57.45 | 42.55 | 89.30 | 10.70 |
| July 2010 | 99.72 | 0.28 | 58.28 | 41.72 |
| Aug. 2010 | 64.62 | 35.38 | 34.91 | 65.09 |
| Sept. 2010 | 38.69 | 61.31 | 31.23 | 68.77 |
| June 2011 | 67.67 | 32.33 | 67.56 | 32.44 |
| July 2011 | 88.75 | 11.25 | 62.06 | 37.94 |
| Aug. 2011 | 89.88 | 10.12 | 64.86 | 35.14 |

The proportion of the new water during the three wet seasons in both catchments is more towards the rising limb of the hydrograph. This implies that the new water component is generated via surface hydrological flow pathways in both of the catchments, and has a greater proportion than the pre-event component. Surface runoff generation starts immediately after the rainfall event in agricultural fields, in grazing lands and on bare lands. The existing gully formation as a result of severe erosion from different land use also corroborates the results of the isotope study, which shows the dominance of the event water proportion in these catchments. This is also supported by the observed flashy behaviour of small catchments (Temesgen et al., 2012). However, it is noted that the pre-event water dominates after the main event water peak.

Furthermore, it can be assumed that the high percentage of event water in both catchments is due to the low infiltration rate and the compaction of the top soil in the agricultural lands. Research in the vicinity of these catchments also suggests that the effect of a plough pan due to long years of ploughing activities reduces the infiltration capacity of the soil (Temesgen et al., 2012). The effect of topography, soil physical parameters and land degradation could also be another factor in the large proportion of the event water component (Teferi et al., 2013). Nonetheless, during winter (dry season) and spring (small rainy season) the river water at both catchments is solely derived from the groundwater recharged during the rainy (summer) season. Similar studies in the US from small agriculturally dominated catchments showed that event water has a large proportion of runoff components due to low infiltration rates of agriculturally compacted soils (Shanley et al., 2002).

Past studies in different regions showed that the pre-event water is the dominant runoff component during the event (e.g. Sklash and Farvolden, 1979; Pearce et al., 1986; McDonnell, 1990; Mul et al., 2008; Hrachowitz et al., 2011; Munyaneza et al., 2012). However, the results from the present study showed that event water is the dominant runoff component. The possible reasons could be an agriculturally dominated catchment with a high soil erosion affected area, steep slopes (e.g. varying between 2% in the lower part of the catchment and more than 45% in the upper part) and high seasonality of climate; the event water proportion is the dominant runoff component during the wet season on the seasonal timescale. Thus, due to these factors, the event water proportion is the dominant runoff component during the wet season on the seasonal timescale.

### 6.4.5 Uncertainty analysis of the hydrograph separations

Table 6.5 presents the results of the uncertainty analysis for the seasonal hydrograph separation at the 70% (approximately one standard deviation) confidence interval. The computed total uncertainty is based on the isotope concentrations in the month of February (low flow) as pre-event end member for the whole wet season. The uncertainty results from this end member concentration are relatively low compared to the sensitivity made for the different pre-event end member concentrations.

The average uncertainty terms arising from the pre-event, event and river water for the three wet season periods accounted for 7, 61 and 32% for the Chemoga catchment and 4, 51 and 45% for the Jedeb catchment. This is suggests that most of the uncertainty stems from the event water component. Genereux (1994) pointed out that the greater uncertainty can mainly be attributed to the proportions that contribute the higher runoff components.

The error in hydrograph separation originates from different sources (see Uhlenbrook and Hoeg, 2003 for details). Based on the uncertainty results, the fraction of the different components using different pre-event end member concentrations gives only a range of values, not the exact number. Thus, due to spatial and temporal variation in the end member concentrations, the classical hydrograph separations methods give only a qualitative description of the runoff components and their variable contributions in time (Uhlenbrook and Hoeg, 2003).

Table 6.5 Percentage of total uncertainty in event, pre-event and stream water concentrations during the wet season in the Chemoga and Jedeb catchments.

| Month | Chemoga catchment | | | Jedeb catchment | | |
|-------|---------|---------|------|---------|---------|------|
| | $Q_E$ (%) | $Q_{Pe}$ (%) | Q (%) | $Q_E$ (%) | $Q_{Pe}$ (%) | Q (%) |
| July 2009 | 96.90 | 0.02 | 3.08 | 96.3 | 0.00 | 3.69 |
| Aug. 2009 | 54.21 | 2.67 | 43.12 | 25.6 | 3.34 | 71.05 |
| Sept. 2009 | 41.37 | 58.63 | 0.00 | 58.7 | 1.75 | 39.51 |
| June 2010 | 14.41 | 7.35 | 78.24 | 76.2 | 0.13 | 23.64 |
| July 2010 | 72.80 | 0.00 | 27.20 | 55.9 | 3.35 | 40.72 |
| Aug. 2010 | 94.67 | 2.96 | 2.37 | 44.5 | 9.25 | 46.23 |
| Sept. 2010 | 11.00 | 10.21 | 78.79 | 29.2 | 12.93 | 57.92 |
| June 2011 | 82.86 | 14.03 | 3.11 | 36.6 | 3.00 | 60.43 |
| July 2011 | 67.72 | 0.29 | 31.99 | 39.8 | 3.83 | 56.39 |
| Aug. 2011 | 72.22 | 0.01 | 27.77 | 42.7 | 3.16 | 54.15 |
| Average | 61.00 | 9.00 | 30.00 | 51.00 | 4.00 | 45.00 |

**6.4.6 Estimation of mean residence times**

Figure 6.14 presents precipitation and streamflow seasonal in $\delta^{18}O$ patterns for the estimation of the mean residence time of water based on the periodic regression analysis to fit the seasonal sine wave models. Preliminary estimation of mean residence time was obtained using the model described in equations 6.9 - 6.11 and results are provided in table 6.6. Based on the seasonal variation in $\delta^{18}O$ both in precipitation and streamflow, the mean residence times are estimated as 4.1 and 6.0 months in the Chemoga and Jedeb catchments, respectively. The goodness of fit of the observed streamflow output isotope signal is moderate as, is shown by the coefficients of determination ($R^2$ varies from 0.47 to 0.66).

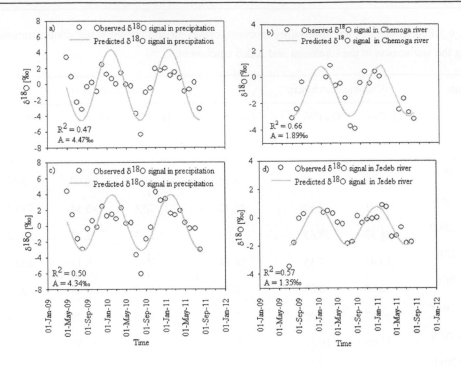

Figure 6.14: Fitted sine wave regression models to $\delta^{18}O$ values for precipitation and river water a) and b) in the Chemoga, and c) and d) in the Jedeb river. Inside the figures, $R^2$ is the coefficient of determination, and 'A' is the amplitude for the input and output isotope signals.

The method is appropriate for the short record length and coarse frequency of spatial and temporal tracer sampling. Indeed, the method gives indicative first approximation estimates of mean residence times and the level of fit is in line with previous studies (Soulsby et al., 2000; Rodgers et al., 2005a). The results of short mean residence times in both catchments are in line with the hydrograph separations, which indicate more surface runoff generation than base flow contribution during the storm events in these steep headwater catchments.

Table 6.6 Amount-weighted mean precipitation and flow $\delta^{18}O$ composition, estimated amplitude, phase lag $\Phi$ and mean residence time in the Chemoga and Jedeb catchments over the period July 2009-Aug 2011.

| Description | Mean annual measured $\delta^{18}O$ [‰,VSMOW] | Amplitude [‰,VSMOW] | Phase lag $\Phi$ [radian] | Mean residence time [months] |
|---|---|---|---|---|
| Chemoga catchment precipitation | -0.58 | 4.47 | 1.11 | |
| Chemoga discharge | -1.34 | 1.89 | 0.01 | 4.1 |
| Jedeb catchment precipitation | 0.40 | 4.34 | 1.12 | |
| Jedeb discharge | -0.70 | 1.35 | 0.01 | 6.0 |

The results of preliminary estimation of mean residence time are plausible and anticipated from steep agriculturally dominated catchments with little adoption of soil and water conservation measures, which enhance more surface runoff generation (Temesgen et al., 2012). Consequently, the surface condition or the responsiveness of the soil due to the plough pan effect influences the ability of the soil to infiltrate the given rainfall amount to recharge the groundwater system (Tekleab et al., 2014a). Nonetheless, this result cannot be generalised to other regions with agriculturally dominated catchments. Other factors like the soil infiltration and retention conditions, slope, drainage network and other parameters and processes could also alter/influence the runoff generation mechanism. However, in these case study catchments, the overland flow is the dominant runoff component.

Furthermore, study in the north-western Ethiopian plateau reported that the groundwater in the area is characterised by shallow and rapid circulation leading to a young age of the groundwater system, which might be the cause of the drying out of groundwater wells after prolonged droughts (Kebede, 2004). However, the groundwater age distribution is not yet fully understood in the area, and needs further research.

The mean residence time in a catchment varies depending on topography, soil types, land cover, and geologic properties (McGlynn et al., 2003; Tetzlaff, et al., 2007a and 2009). The estimated mean residence times in this study are in agreement with similar meso-scale catchments around the world. For instance, Rodgers et al. (2005a) found a residence time of 6.8 months in the Scotland Feugh meso-scale catchment, which is dominated by responsive soils. In a similar isotope study in Scotland in a nested meso-scale catchment (the catchment area varies from 10 to 231 km$^2$), the estimated mean residence time varies from 1 month to 14 months (Rodgers et al., 2005b) Uhlenbrook et al. (2002) found the residence times of 24 to 36 months for the shallow groundwater, and 6 to 9 years for the deep ground water in the Brugga (40 km$^2$) black forest meso-scale catchment in Germany. Although the climate, topography, land use, soil and geology of the catchments in the present study are different from those of other investigated catchments, the estimated mean residence times are comparable.

It is apparent from the above discussion that the mean residence time is not directly dependent on the catchment size. However, the mean residence times might be longer for small headwater catchments (McGlynn et al., 2003). It is also noted that the mean residence time of a catchment is influenced by the heterogeneity in climatic setting, topography and geology (Hrachowitz et al. 2009), landscape controls, particularly soil cover (Soulsby, et al. 2006), percent coverage responsive soil using catchment soil maps (Soulsby and Tetzlaff, 2008), and topography and soil drainage conditions in different geomorphic provinces (Tetzlaff et al., 2009). A recent study by Heidbüchel et al. (2012) showed that the mean residence time is better estimated using time-variable functions. Thus, the residence time is depends not only on the catchment characteristics but also on time varying climate inputs. Nevertheless, investigating all these different controlling processes on mean residence time estimation was beyond the scope of the present study. Therefore, there is a need for further research that takes the influence of catchment heterogeneity on the estimation of mean residence time into account in catchments of the Nile basin.

## 6.5 Conclusions

Characterisation of stable isotope composition of precipitation, spring and river water along different altitude gradients were undertaken with the aim of preliminarily estimating the mean residence time and runoff component contributions on a seasonal timescale. The results show that precipitation, stream and spring waters exhibit noticeable spatial and temporal variations in stable $\delta^{18}O$ and $\delta^2H$ composition in the study area.

The results further demonstrate that the meteoric water in the study area is influenced by the amount and to a lesser extent by the altitude and temperature effects. The climatic seasonality, which is dominated by different moisture sources, along with the local meteorological settings plays a significant role in the isotopic composition of rainfall in the area.

The analyses of isotope results reveal the dominance of event water and short mean residence times in both of the catchments. From the point of view of managing the water resources and the importance of the available soil water for consumptive use of the crops, catchment management aiming at reducing overland flow/soil erosion and increasing soil moisture storage and recharge has paramount importance for the farmers residing in these catchments.

It should be noted that in the light of the data availability the estimated mean residence times and seasonal hydrograph separation represent first approximations. Consequently, for more reliable estimates of the mean residence times and runoff contributions there is a need for further research with finer resolution sampling during storm events and long-term isotope tracer data collection at different spatial and finer temporal scales (e.g. daily and hourly) that will improve our understanding of how these catchments function. It is noteworthy that the applied methods were used for the first time in the region that has critical regional importance regarding the water resources in the Nile. Thus, the results can be used as a baseline for further hydrological studies for a better understanding of the dominant runoff components in the future.

# CHAPTER 7

## Catchment modelling through the use of Stable Isotope data and field observations in the Chemoga and Jedeb meso-scale catchments, Abay/Upper Blue Nile basin, Ethiopia[1]

The previous chapters provided trend analysis of hydro-climatic variables, water balance investigation, hydrologic responses of a catchment to land use change, and use of stable isotopes for a better understanding of the runoff components and estimation of mean residence time in the meso-scale catchments. To gain insights about runoff generation processes in headwater meso-scale catchments intensive data collection of daily precipitation, water levels, and stable environmental isotopes have been done. The objective of this study is to understand the rainfall-runoff processes of the meso-scale Chemoga and Jedeb catchments in the Abay/Upper Blue Nile basin, Ethiopia. Distributed conceptual model is developed in PCRaster software modelling environment. Three different model representations with varying model complexity have been employed to test the appropriate model structure. Parameters were conditioned within the Generalized Likelihood Uncertainty Estimation (GLUE) framework using both discharge and environmental isotope information indicating the ratio of new and old water during the wet season. It has been demonstrated that the two catchments cannot be modelled equally well with the same model structure due to differences in the rainfall-runoff processes caused by different amount of wetlands that is leading to different hydrological responses . Hence a single lumped model structure for the entire Abay/Upper Blue Nile cannot do justice to all dominant hydrological processes in the various sub-catchments in the basin.

## 7.1 Introduction

Understanding rainfall-runoff processes within the catchments of the Abay/upper Blue Nile is crucial for sustainable management of the limited water resources in the basin (Kim and Kaluarachchi, 2008; Steenhuis et al., 2009). Hydrological models are increasingly important to investigate the hydrological system dynamics and to predict the potential impacts of changes in the catchment on the discharge regime, at various spatial and temporal scales (Wagener et al., 2001; Buytaert and Beven, 2011). However, developing hydrological models to support management of real world problems is a challenging task in hydrology (Kirchner, 2006; Savenije, 2010).

---

[1] *This chapter is based on*: Tekleab, S., Uhlenbrook, S. Savenije, H.H.G., Mohamed, Y., and Wenninger, J. (2014). Catchment Modelling through the use of stable environmental isotopes and field observations in the Chemoga and Jedeb meso-scale catchments, Abay/Upper Blue Nile basin, Hydrological Science Journal (Under review).

On the one hand, observations of climatic and hydrometric data sets are scarce, in particular in the developing world, and subject to measurement errors, therefore, it is problematic to study hydrological processes and their relationship to catchment characteristics at the right spatio-temporal scales based on observation. On the other hand, existing hydrological models are often inadequate to simulate the dominant hydrological processes owing to parameter and model structural uncertainties (Uhlenbrook et al. 1999; Sivakumar, 2008; Savenije, 2009; 2010; Clark et al., 2011a).

The literature covers different approaches to conceptualise and model dominant hydrological processes within a flexible modelling framework. Clark et al. (2008; 2011a) tried to identify the most appropriate model using a range of existing model structures. Fenicia et al. (2007; 2008a) introduced a fully flexible model structure to understand catchment behaviour. Savenije (2010) suggested making use of hydrological landscape classification to identify model structures that best describes the dominant hydrological processes in a catchment. Clark et al. (2011b) and Buytaert and Beven (2011) formulated multiple working hypotheses to test the appropriate representation of hydrological systems. However, with these large number of published work, understanding catchment processes still need the development of low parameterised model with less predictive uncertainty (Savenije, 2001; Sivapalan, 2009).

The uncertainties stemming from different sources, viz, forcing data model parameters and model structure exert additional challenges for predicting hydrological impacts of changes in land use and climate. A large number of publications are already available addressing hydrological model development and uncertainty analysis, (e.g., Beven and Binley, 1992; Uhlenbrook et al., 1999; Sivapalan et al., 2003c; Wagener et al., 2003; Vrugt et al., 2003; Beven, 2008; Sivapalan, 2009), among others. However, the literature shows limited applications in data scare environments (Winsemius et al., 2009).

Therefore, constraining model parameters using auxiliary data, such as stable environmental isotopes, in addition to streamflow data, has the potential to increase the degree of trust in model prediction (Seibert and McDonnell, 2002; Wissmeier and Uhlenbrook, 2007). Forinstance, Winsemius et al. (2009) conditioned the model parameters by defining limits of acceptability within the Generalized Likelihood Uncertainty Estimation (GLUE) framework. They have implemented a distributed conceptual model in the data scare Luangwa basin in Zambia, using soft information such as expert judgment, and hard information in the form of observed hydrological signatures (based on hydrograph analysis). In particular, data from the recession characteristics of flow and spectral properties of discharge series as a target values were considered as hard information. Son and Sivapalan (2007) used stable environmental isotope data in conjunction with groundwater level measurements for improving the model structure, parameter identifiability, and reduced of predictive uncertainty for the Susannah Brook catchment in Western Australia.

There exists a considerable literature on the hydrology of the Abay/Upper Blue Nile basin, ranging from simple water balance models on a monthly and daily time scale (see e.g. Conway, 1997; Kebede et al., 2006; Kim and Kaluarachchi, 2008; Steenhuis et al., 2009; Uhlenbrook et al., 2010). More complex model such as the Soil and Water Assessment Tool (SWAT) model was applied in Lake Tana sub-basin (Setegne et al., 2010). Easton et al. (2010) and White et al. (2010) developed the modified SWAT-WB water balance model that takes in to account the saturation excess runoff routine for predicting both flow and sediment in the Blue Nile basin. The findings of these studies showed that overland flow occurs most often at the bottom of the hillsides. However, most of those studies were conducted on large scale to analyze the flow at the outlet at the Ethiopia-Sudan border. Some of the studies are limited to Lake Tana sub-basins with limited spatial coverage of precipitation data. (Haile, 2010).

Moreover, no auxiliary data have been used to support model construction, calibration or validation. Collick et al. (2009) used a semi-distributed water balance model based on the Thornthwaite and Mather, (1955) that has been applied in the Soil Conservation Research Project (SCRP) in small catchments Anjeni (1.13 km$^2$), Andit Tid (4.8 km$^2$) and Mayber (1.13 km$^2$). They reported that with similar calibrated parameters used in small catchments, the model is able to predict flow well reasonably with equal Nash-Sutcliffe efficiency at the larger Blue Nile scale and the scale effect is minimal. However, hydrological processes are heterogeneous at all spatial and temporal scales (e.g. Blöschl and Sivapalan, 1995), and linking the processes in scaling relationships are a key to identify process controls on the appropriate spatio-temporal scales (Didszun and Uhlenbrook, 2008).

Recently, Tilahun et al. (2013,a,b; 2014) developed a coupled flow and sediment model that has been tested in small experimental catchments within the upper blue Nile basin, sizes ranging from 0.95 km$^2$ at Debre Mawi to 4.8 km$^2$ at Andit Tid catchment and at larger scale in the Upper Blue Nile 174,000 km$^2$. They assigned model to use the landscape as different runoff generation units. In the hillside degraded area, direct runoff is conceptualized, in the middle hillside infiltration zone, recharge, inter flow and base flow is conceptualized and saturated overland flow was assumed to be occurred at saturated bottom area at the foot of the hillslopes near to the river. Their results showed that the model reproduces the observation well and reported that saturation excess runoff is the dominant mechanism both at the smaller scale and at the larger basin scale.

The weakness of these past researches in the upper Blue Nile is that the prediction of flow is evaluated using a single objective measures i.e. the Nash-Sutcliffe efficiency ($E_{NS}$). Hence, the information content in the hydrographs was explored only during the highflows. Furthermore, no techniques have been employed to handle equifinality in a feasible parameter space using methods of uncertainty assessment (e.g. Beven and Binley, 1992; Vrugt et al., 2003).

The present paper focuses on investigating the rainfall–runoff processes of Chemoga and Jedeb meso-scale catchments. Unlike the previous studies, this paper use a combination of hydro-metric, stable isotope data and field process knowledge to constrain model parameters during calibration process in a multi-objective sense and parameter uncertainty was assessed using the (GLUE) methodology. The catchments are headwater catchments in the Abay/upper Blue Nile basin; both are mountainous and dominated by agricultural land use.

The objectives of this study are: (1) to identify the dominant hydrological processes in these catchments, and (2) to demonstrate the value of isotope data in constraining model parameters and in reducing predictive uncertainty.

## 7.2. Data source

The geographical location, land use, soil, and geological characteristics of Chemoga and Jedeb catchments are described in chapter 6. Here the data sets used for modelling is described in subsequent section.

### 7.2.1 The rating curve

Streamflow data set based on manual water level measurements were taken (daily at 06:00 a.m. and 06:00 p.m.) beginning from 01/07/2009-31/07/2012 at the outlet of Chemoga and Jedeb gauging stations, respectively. Stage discharge (Q-h), relationships were then developed to change the water level into discharge time series using the least square regression models.

The rating curve equation is given by equation 7.1.

$$Q = a(h - h_0)^b \tag{7.1}$$

Where, Q is the discharge ($L^3\ T^{-3}$), h is the water level (L), $h_0$ is the water level (L) corresponding to no discharge in the river, a [$L^{3-b}\ T^{-1}$] and b [ - ] are rating constants.

Current meter device has been used to measure the velocity of flow. The discharges in the rivers were measured using area velocity method between 5/08/2011 until 2/09/2011 (see fig. 7.1). A total of 16 and 14 rating curve points have been measured in the main rainy season in Chemoga and Jedeb Rivers, respectively[2]. Equation 7.1 is compatible with the Manning formula (equation 7.2). The geometric properties, cross-sectional area and hydraulic radius in equation 7.2 are a function of (h-$h_0$) in equation 7.1.

---

[2] *Acknowledgement to Mr. Birhanu Legesse, and Mr. Beyene Minda, who are employee of Ministry of Water and Energy, Ethiopia for giving technical support in measuring the velocity of flow.*

$$Q = \frac{A}{n}R^{2/3}S^{1/2} \tag{7.2}$$

Where A, is the cross-sectional area [L], n, is the Manning roughness coefficient [T $L^{-1/3}$], R, is the hydraulic radius of the channel [L], and S is the slope of the channel [-]. As shown in fig. 7.1, the measured discharge was limited to the extent that, it does not cover a wide range of discharge values. Thus, the measured discharges using Stevens (1907) method were extrapolated to the desired water level. In fig. 7.2 the first part (the curve), is derived from a topographical survey of the cross-section. This curve can be extrapolated as far as necessary, as high as the water level may go. The other part, the line, requires a couple of discharge measurements, but because it is a straight line, it requires only a limited number of points and subsequently we can use that line to combine with the curve to obtain the extrapolated discharge at the desired water level. However as depicted in figure 7.1, the over bank flow cannot be estimated by the Stevens method, since the cross-sectional area could not go beyound the bank and measurements were not taken.

From the least square solution using equation 7.1, the values of a, b and $h_0$ for the Chemoga river found to be 14.13, 1.83, and 0.1, respectively. In the Jedeb catchment, the values of a, b and $h_0$ was estimated to 19.01, 2.12, and 1.05, respectively. However, discharge values are better extrapolated using the Stevens method due to the fact that the water level and the discharge are related to the cross-sectional area mathematically through ($AR^{2/3} = f(h\text{-}ho)$) and ($Q = g(AR^{2/3})$). Equation 7.1 can be related to the regression equations in figure 7.2 by taking the same ho value. Thus, the method separates the geometry of the cross-section from its counterparts, the roughness and slope ($S^{1/2}/n$) (see figure 7.2).

Figure 7.1: Velocity measurements for development of rating curves in Chemoga left and Jedeb rivers right during high flow period (Photo taken by the author August 2011).

The climatic and hydrological variables mentioned below computed over the period 2009-2011 hydrological year. The annual rainfall ranges between1300-1600 mm $a^{-1}$ in the lower and upper part of the catchments. The mean annual average daily temperature computed using Thiessen polygon method from four stations are 13.9°C in Chemoga and 15°C in the Jedeb catchment, respectively. The average annual potential evaporation computed using Hargreaves method amounts 1340 mm a$^{-1}$ in the Chemoga and 1390 mm a$^{-1}$ in the Jedeb catchments (Hargreaves and Samani, 1982). The mean annual discharge in Chemoga and Jedeb catchment is 6.6 and 7.9 m$^3$ s$^{-1}$, respectively.

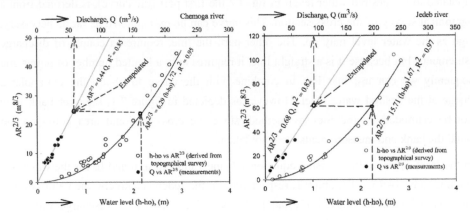

Figure 7.2: Extrapolation of rating curves for Chemoga and Jedeb rivers.

## 7.3. Methodology

### 7.3.1 Model setup

A lumped conceptual distributed model has been used to understand the rainfall-runoff processes of Chemoga and Jedeb catchments. The model structures were programmed with the PCRaster software modelling software (Wesseling et al., 1996) with a grid size of 200*200m$^2$. PCRaster modelling offers a programming environment for distributed modelling (e.g., Karssenberg et al., 2001; Ott and Uhlenbrook, 2004; Wissmeier and Uhlenbrook, 2007). It offers also the opportunity to account for spatial heterogeneities of land use, soils and topography. The spatial maps are combined using a dynamic modelling script to represent the different process description of the hydrological system.

PCRaster uses the local drainage direction map extracted from the digital elevation map to simulated lateral flows. At each time step in each grid cell the surface runoff accumulated through local drainage direction generated the river flow without an additional routing routine. Note that the modelling time step is daily and it is assumed that the runoff concentration processes are faster than the modelling time that had to be applied due to data availability. An additional routing routine would have increased the model complexity and, thus, the number of parameters and associated model uncertainty. The model structure used is a modification of the lumped elementary watershed model applied in the upper Zambezi river basin (Fenicia et al., 2006; Winsemius, 2009). The model structures are schematized in figure 7.3.

The storage compartments consist of an interception storage $S_i$ [L], soil moisture storage in the unsaturated zone $S_u$ [L], fast flow generating storage $S_f$ [L] and two slow flow generating storage; the saturated storage $S_s$ [L] and the wetland storage $S_w$ [L] conceptualize the processes in Chemoga catchment, while in Jedeb only saturated storage (without wetland storage) is used.

The model used distributed inputs of precipitation, potential evaporation and soil moisture states. However, all the parameters are lumped. In Chemoga the wetland is modelled with the assumption that the fluxes from unsaturated storage first fill the saturated storage until a threshold depth $S_{s,th}$ [L T$^{-1}$] is exceeded. The flux above this threshold depth is spilled into the storage of the wetland defined by area of wetland ($A_w$) (see fig.7.3). This implies that the recharge from the hillslopes feed the groundwater as a result of which the saturation in the wetland expands. The wetland storage also has a threshold depth $S_{w,th}$ [L T$^{-1}$]; when this threshold is exceeded the fast discharge is modelled according to linear storage discharge relationship. At a daily time scale interception can be modelled by a simple threshold process, whereby the evaporation from interception is less than minimum of the daily interception capacity, the potential evaporation and the daily rainfall (de Groen and Savenije, 2006).

$$E_i = \min(P, I_{max}, E_p)$$
(7.3)

Where $E_i$ [L T$^{-1}$] is the interception evaporation $P$ is the precipitation [L T$^{-1}$], $I_{max}$ [L T$^{-1}$] is the daily interception capacity and $E_p$ is the potential evaporation [L T$^{-1}$]. The precipitation exceeding the maximum threshold for interception is the effective precipitation $P_e$ [L T$^{-1}$], partitioned into different components after infiltrating into the unsaturated soil moisture reservoir $S_u$ [L]. The flux in unsaturated soil $R_u$ [L T$^{-1}$] is the water infiltrated into the unsaturated reservoir, and $R_s$ [L T$^{-1}$] is the flux entering into the fast reservoir $S_f$ [L], when the maximum soil moisture capacity of the unsaturated reservoir is exceeded. The flux $P_r$ [L T$^{-1}$] is the preferential recharge directed towards the saturated reservoir $S_s$ [L] which is computed based on the runoff generation coefficient $R_c$ [-] and the runoff partitioning coefficient $\alpha$ [-]. $R_c$ [-] is runoff generation coefficient modelled based on Fenicia et al. (2006) represented by S shaped logistic function, which relates the soil moisture state $S_u$ [L] and maximum soil moisture capacity $S_{max}$ [L] to the shape factor $\beta$ [-]. $\beta$ [-], describes the spatial distribution of the soil moisture in the catchment.

$$R_c = \frac{1}{1 + \exp\left(\dfrac{-S_u / S_{max} + 1/2}{\beta}\right)}$$
(7.4)

$$R_u = (1 - R_c)$$
(7.5)

$$R_s = \alpha R_c P_e \tag{7.6}$$

Where $\alpha$ [-] is the partitioning coefficient for fast surface runoff and preferential recharge.

$$P_r = (1 - \alpha)(R_c P_e) \tag{7.7}$$

The percolated water $P_c$ [L T$^{-1}$] from the unsaturated reservoir $S_u$ [L] in to saturated

reservoir $S_s$ [L] is computed using linear relationship of the soil moisture with the maximum

percolation rate $P_{max}$ [L T$^{-1}$] computed using eqn. 7.8 (Fenicia et al., 2006).

$$P_c = P_{max}\left(\frac{S_u}{S_{max}}\right) \tag{7.8}$$

The actual evaporation from the unsaturated reservoir $T_a$ [L T$^{-1}$] is computed according to the following formula

$$T_P = \max(E_p - E_i, 0) \tag{7.9}$$

$$T_a = \min\left(\frac{S_u}{L_p S_{max}}\right) T_P \tag{7.10}$$

Where $T_P$ [L T$^{-1}$] is the potential transpiration and $L_P$ is the fraction of $S_{max}$ below which $T_P$

is constrained by $S_u$.

The flux $Q_f$ from fast reservoir $S_f$ and $Q_s$ from slow saturated reservoir $S_s$ modelled as

linear reservoir storage discharge relationship according to equation 7.11 and 7.12.

$$Q_f = K_f S_f \tag{7.11}$$

$$Q_s = K_s S_s \tag{7.12}$$

Where $K_f$ [T$^{-1}$] is the fast reservoir coefficient and $K_s$ [T$^{-1}$] is the slow reservoir coefficient.

The observed peak flows in the Jedeb catchment are modelled by inclusion of a threshold depth for the volume of water exceeded the maximum capacity of fast reservoir $S_{f,max}$ for overland flow generation, $Q_{OF}$ [L$^3$ T$^{-1}$], computed using the following formula.

$$Q_{OF} = \max(S_f - S_{f,max}, 0) \tag{7.13}$$

The descriptions of model parameters and their prior range are shown in table 7.1 and the

main differences between the model versions $M_1$, $M_2$, and $M_3$ in both catchments are defined

in table 7.2. In the table 7.1,"es" denotes parameters are estimated, "n/a" denotes parameter

not used in the respective catchment model, and $\sqrt{}$ indicates that parameters are used and

determined during calibration. The maximum interception threshold is fixed as 2 mm/d for the

dominant land use type of agriculture. This value is assumed to be adequate for most of the river basin in Africa (de Groen and Savenije, 2006). Time scales for the slow reservoir parameter were determined from the master recession curve as 0.183 [1/d] and 0.204 [1/d] for Chemoga and Jedeb catchments, respectively.

Table 7.1: Description of the model parameters and their prior range

| S. No. | Parameter [units] | Chemoga | Jedeb | Parameter description | Lower and upper limit |
|---|---|---|---|---|---|
| 1 | $I_{max}$ [L T$^{-1}$] | Es | es | Maximum interception threshold | Literature |
| 2 | $\beta$ [-] | √ | √ | Shape parameter for runoff generation | [1, 5] |
| 3 | C[L T$^{-1}$] | √ | √ | Maximum water uptake rate by plants | [0.01, 0.4] |
| 4 | $L_p$ [-] | √ | √ | Fraction of maximum soil moisture capacity constraining transpiration | 0.5 (assumed) |
| 5 | $\alpha$ [-] | √ | √ | Runoff partitioning coefficient | [0, 1] |
| 6 | $K_s$ [T$^{-1}$] | Es | es | Time scale for slow reservoir | Recession curve |
| 7 | $K_f$ [T$^{-1}$] | √ | √ | Time scale for fast reservoir | [0.01, 1.8] |
| 8 | $P_{max}$ [L T$^{-1}$] | √ | √ | Maximum percolation rate | [0.01, 0.3] |
| 9 | $S_{max}$ [L] | √ | √ | Maximum soil moisture capacity | [50, 400] |
| 10 | $S_{s,th}$ [L] | √ | n/a | Threshold for saturated reservoir | [1,50] |
| 11 | $S_{w,th}$ [L] | √ | n/a | Threshold for the wet land | [1.5, 10] |
| 12 | $S_{F,max}$ [L] | n/a | √ | Threshold for infiltration excess over land flow | [0.1, 5] |

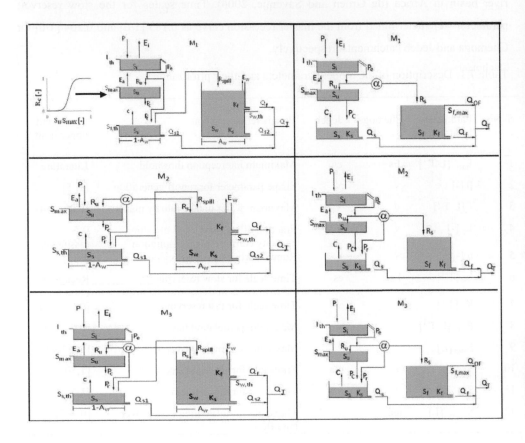

Figure 7.3: Model structures used for the studied catchment. The left side represents the Chemoga catchment and the right side represents Jedeb catchment.

Table 7.2: Description of the different model representations in the Chemoga and Jedeb catchment.

| Catchment | Model | Assumptions/model architecture, dominant runoff process | Model parameters conditioned/calibration | Conceptualisation |
|---|---|---|---|---|
| Chemoga | $M_1$ | $\alpha^{**}$ has a value of zero, dominant rapid sub-surface flow | Rainfall-runoff data alone | Interception component, fast runoff entering to fast reservoir, no fast over land flow component exceeded maximum capacity of fast reservoir |
|  | $M_2$ | $\alpha$ is used to partition the runoff in to surface and sub-surface flow | Rainfall-runoff data and stable isotope data (wet season percentages of new and old water) | No interception component, fast runoff entering to fast reservoir, no fast over land flow component exceeded maximum capacity of fast reservoir |
|  | $M_3$ | $\alpha$ is used to partition the runoff in to surface and sub-surface flow | Rainfall-runoff data and stable isotope data (wet season percentages of new and old water) | Interception component, fast runoff entering to fast reservoir, no fast over land flow component exceeded maximum capacity of fast reservoir |
| Jedeb | $M_1$ | $\alpha$ has a value of 1, i.e. surface runoff is dominant, no rapid sub-surface flow | Rainfall-runoff data alone | Interception component, fast over land flow component exceeded maximum capacity of fast reservoir |
|  | $M_2$ | $\alpha$ is used to partition the runoff in to surface and sub-surface flow | Rainfall-runoff data and stable isotope data (wet season percentages of new and old water) | Interception component, no fast over land flow component exceeded maximum capacity of fast reservoir |
|  | $M_3$ | $\alpha$ is used to partition the runoff in to surface and sub-surface flow | Rainfall-runoff data and stable isotope data (wet season percentage of new and old water) | Interception component, fast over land flow component exceeded maximum capacity of fast reservoir |

$^{**}$ The runoff partitioning coefficient defined in table one.

## 7.3.2 Calibration of parameters

Manual calibration has been used initially by adjusting parameters in an iterative way until the observed system output exhibited an acceptable level of agreement with model output based on the Nash-Sutcliffe efficiency ($E_{NS}$) greater than 0.7 for daily simulation. This helped to fix the range of parameters for automatic calibration and to distinguish preliminary sensitive model parameters, which reduce the calibration time during automatic calibration. In a second step, multi-objective calibration following the method by Gupta et al. (1998) and Boyle et al. (2000) has been used to calibrate model acknowledging the different characteristics of the hydrograph during high flows and low flows. The Monte Carlo Analysis Tool (MCAT; Wagener et al., 2003) has been used for calibrating the parameters in a feasible parameter space. The feasible parameter space is defined based on past research experience within the basin e.g. Uhlenbrook et al. (2010), expert judgment from field observation and preliminary results of manual calibration. Uniform random sample of 10,000 Monte Carlo (MC) run was implemented using MCAT. Two objective functions based on Nash and Sutcliffe (1970) were used: the Nash Sutcliffe efficiency of the discharge and the Nash Sutcliffe efficiency using logarithmic discharge values, with the aim to give equal weight high and low flows described in equation 7.14 and 7.15.

$$F_{HF} = \frac{\sum_{i=1}^{n}(Q_{sim,i} - Q_{obs,i})^2}{\sum_{i=1}^{n}(Q_{obs,i} - \overline{Q}_{obs,i})^2} \qquad (7.14)$$

$$F_{LF} = \frac{\sum_{i=1}^{n}[\ln(Q_{sim,i}) - \ln(Q_{obs,i})]^2}{\sum_{i=1}^{n}[\ln(Q_{obs,i}) - \ln(\overline{Q}_{obs,i})]^2} \qquad (7.15)$$

Where $Q_{sim,\ i}$ [m$^3$ s$^{-1}$] is the simulated streamflow at time $i$, $Q_{obs,\ i}$ [m$^3$ s$^{-1}$] is the observed streamflow at time $i$, $n$ is the number of time steps in the calibration period and the over-bar indicates the mean of observed streamflow. The objective function $F_{HF}$ was selected to minimize the errors during high flows (same formula as $E_{NS}$), and $F_{LF}$ uses logarithmic values of streamflow and improves the assessment of the low flows.

### 7.3.3 Constraining model parameters using isotope data

The usefulness of additional auxiliary data to gain confidence in the model prediction has been demonstrated by many authors (e.g. Seibert and McDonnell, 2002; Sieber and Uhlenbrook, 2005; Son and Sivapalan, 2007; Fenicia et al., 2008b; Winsemius et al., 2009). In this study isotope information has been used to further constrain the model parameters. Results from a two component mixing model for separating the hydrograph at the seasonal time scale have been used (see Tekleab et al., 2014b). The flow chart describing the methodology is shown in figure 7.4. The methodology described above was implemented in a Monte Carlo experiment within MCAT (Wagener et al. 2003) employing the following steps:

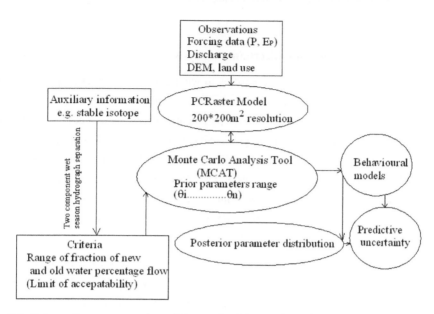

Figure 7.4: Schematic representation of the methodology adopted for calibration and analysis of predictive uncertainty.

(i) The fraction of new (event) water and old (pre-event) water flows are set as criteria to constrain the model parameters in both catchments. From the seasonal hydrograph separation results of Tekleab et al. (2014b) in the Chemoga catchment, the computed event water percentage varies from 62-71% and the pre-event water percentage varies from 29-38%. Whereas in the Jedeb catchment, the event water percentage varies from 52-64% and the pre-event water percentage varies from 36-48%. These relative variations are used to constrain the parameters as limit of acceptability for behavioural models.

(ii) After 10,000 MC ran the simulated daily model outputs are aggregated to monthly discharge values. The corresponding quick flow and slow flow components representing the two fractions of event and pre-event water are also aggregated to monthly discharge values.

(iii) The percentage proportion of quick and slow flows are compared to the event water and pre-event water contributions during wet season (June to September) on a monthly basis and as average proportions during the whole calibration period.

(iv) Any model realization results from the Monte Carlo runs, which satisfy the given average percentage fraction of new and old water within the limits mentioned in step (i) are accepted as behavioural models and any realization outside the limits are considered as non-behavioural models and rejected.

(v) Finally, the retained behavioural models are used for the uncertainty assessment described in section 7.3.4.

### 7.3.4 Parameter sensitivity and uncertainty analysis

Sensitivity analysis of the parameters notifies to what extent the model output is affected by model parameter variations. Generally, insensitive parameters are not well identifiable in the parameter space and they are a sign of over parameterization (Fenicia et al., 2008b). Among the sensitivity analysis methods, the Regional Sensitivity Analysis (RSA) is most commonly used in hydrological modelling studies (e.g. Freer et. al., 1996; Wagener et al., 2001; Wagener et al., 2003; Sieber and Uhlenbrook, 2005). The RSA method is based on uniform random sampling of the parameter space and needs large number of parameter samplings to cover the parameter space. We used the extended RSA, which is implemented in a Monte Carlo Analysis Tool (MCAT) (Wagener et al., 2001; 2003).

The prediction of discharge uncertainty was assessed using the GLUE methodology within MCAT. A Monte Carlo (MC) ran of 10,000 iterations was employed to differentiate between behavioural and non behavioural set of parameters (Freer et al., 1996). For discharge prediction in both catchments (See table 7.2) for the model representation $M_1$, threshold value greater than 0.7 were subjectively chosen to retain the behavioural models. While for the model representation $M_2$, and $M_3$ those parameter sets met the criteria of fraction of new and old water percentage flow described in section 7.3.3 were taken as behavioural models.

The 5-95% range of discharge predictions were computed using the following steps: (i) for each behavioural parameter sets, the likelihood from the MC simulations were arranged in a descending order, (ii) the likelihood values were normalized between 0 and 1 so that it is changed to probability values and the posterior distribution has a cumulative value of one (Beven and Binley, 1992), finally, (iii) for each time step the discharges corresponding to 95% and 5% probabilities from the behavioural models were computed and prediction intervals are constructed.

## 7.4 Results and Discussions

### 7.4.1 Model development

The results of different types of models to represent the hydrological system in both catchments are provided in the following sections.

The results of the hydrograph simulation for the calibration and validation periods are shown in table 7.3. It is illustrated that in both catchments, the model structure $M_1$ gives higher model performance considering the match between observed and simulated discharge hydrograph. However, the model performances are decreasing if the model structure is modified and parameter values are conditioned using stable isotope information. This suggests that high model performance does not necessarily mean that the model comprises a realistic representation of the catchment behaviour (c.f. Savenije, 2001; Sivapalan, 2009). The $M_3$ model representation in both catchments are selected due to the fact that more additional processes are incorporated, parameters are conditioned based on isotope information, parameter identifiability, and less predictive uncertainty.

Table 7.3: Model performance efficiency measures for $M_1$, $M_2$, and $M_3$ representations for the best model in Chemoga and Jedeb catchments in terms of parameters conditioned using rainfall-runoff data alone and stable isotope information using daily simulation.

| Catchment | Model structure | Calibration | | Validation | |
|---|---|---|---|---|---|
| | | $E_{NS}$ [-] | $LogE_{NS}$ [-] | $E_{NS}$ [-] | $LogE_{NS}$ [-] |
| Chemoga | $M_1$ | 0.84 | 0.83 | 0.87 | 0.70 |
| | $M_2$ | 0.70 | 0.63 | 0.78 | 0.62 |
| | $M_3$ | 0.70 | 0.60 | 0.77 | 0.73 |
| Jedeb | $M_1$ | 0.75 | 0.72 | 0.76 | 0.76 |
| | $M_2$ | 0.67 | 0.52 | 0.75 | 0.62 |
| | $M_3$ | 0.73 | 0.53 | 0.79 | 0.65 |

Figure 7.5 presents the observed and simulated discharge results from $M_3$ for both catchments during the calibration and validation periods. It is demonstrated that the peak flows (note the related uncertainty) and the low flows were captured satisfactorily in both catchments. The mismatch of the recessions of the hydrograph at time step 450-500 in the Chemoga catchment could be related to the uncertainties in discharge measurements.

The wet season hydrograph separation results of Tekleab et al. (2014b) in Chemoga and Jedeb catchments support the model architecture of $M_2$ take in to account overland flow process and $M_3$ comprises an interception component as shown in fig 7.3. The calibration results of the retained runoff component based behavioural models depict that almost the same level of model performance achieved constraining the model using isotope data, which later assessed the retained behavioural models based on the objective function towards both high flow and low flow condition. However, during validation the model $M_3$ perform relatively better than $M_2$, which shows the model is good during validation period or in prediction than the calibration period. Previous modelling studies suggest that the explicit inclusion of an interception component in rainfall runoff models improved model efficiency (e.g. Savenije 2004; Zhang and Savenije 2005; Fenicia et al., 2007, Love et al. 2011). Other studies showed reduced model performance (Lindström et al., 1997). In this study, the interception as a dominant process is evaluated not only using model performance but also its role in parameter sensitivity and overall predictive uncertainty (see section 7.4.4).

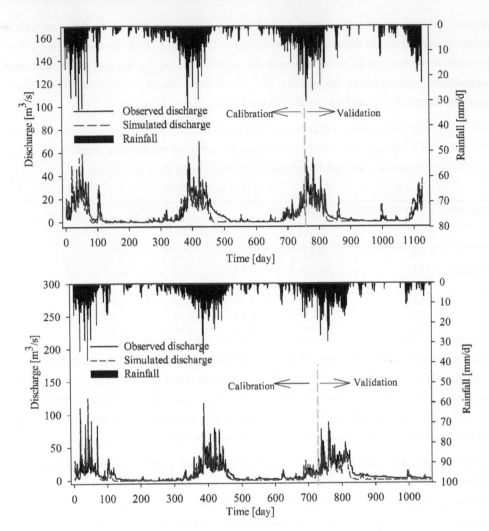

Figure 7.5: Observed and simulated discharge hydrograph during calibration and validation period: Top figure for the Chemoga catchment (model M₃) and lower figure for the Jedeb catchment (model M₃).

The calibration results of model $M_2$ (see fig. 7.3) in the Jedeb catchment characterised by fast responding reservoir without a threshold for the volume of water exceeding maximum capacity of fast reservoir capacity for an overland flow component could not capture the daily streamflow peaks. This model representation was improved by the inclusion of a threshold depth for the volume of water exceeded the maximum capacity of fast reservoir for an overland flow component represented by $M_3$ (see fig. 7.3). Consequently, the peaks of the daily streamflow were captured significantly better by the improved conceptualisation (see fig.7.8f).

Moreover, the result is in line with field observations of overland flow as a dominant runoff component occurring locally in the catchment during events. In these study catchments, due to intensive agricultural practices plough pan effect is prevailing in agricultural fields (Temesgen et al., 2012). As a result the thin layer of degraded soil above the pan, shortly after the rainfall events surface runoff is generated. Hence, due to limited storage capacity, the saturation excess overland flow has occurred. If the rainfall intensity is greater than the infiltration capacity of the soil, infiltration excess over land flow would be generated. However, in the absence of detailed knowledge about rainfall intensities and soil physical characteristics, such a threshold conceptualisation seems reasonable to capture the runoff generation mechanism in a simplified manner.

In terms of the visual inspection and performance measures, all models gave good simulation results. The modelling results in these two catchments highlight the potential of environmental isotope data in refining the initial perception of model representation and to better understand how the hydrologic system is functioning. Moreover, the results using different model representations in these two catchments can be considered as reasonable due the fact that the hydrologic responses of the catchments are different. The long-term annual water balance model results by Tekleab et al. (2011) show that a greater proportion of rainfall received in the Jedeb catchment is released as a quick flow, while greater proportion of water leaves the Chemoga catchment through evaporation due to the presence of extended wetlands.

The existence of wetlands which cover about 10% of the catchment area in Chemoga is likely the reason for their differences in hydrologic responses in particular on the annual water balance. Furthermore, the wetland creates a smoothing effect on runoff response before it reaches to the gauging station and results in lower total observed discharge as compared to Jedeb catchment (see fig. 7.5). Similar conditions within Abay basin show that the neighbouring catchments Gilgel Abay and Koga in the same climatic setting could not be represented by the same model structure, due to the larger extent of wetlands in the later catchment. Thus, transferability of model parameters or prediction at ungauged catchment is challenging (Uhlenbrook et al., 2010). Moreover, within the Upper Abay/Blue Nile basin, study by Kim and Kaluarachchi, (2008) substantiate our results that prediction in the un-gauged basin within the Blue Nile basin is much challenging as the model parameters regionalized in different basin scales were not transferable to one another due to different basin responses.

### 7.4.2 The value of stable environmental isotope data

The model $M_3$ results presented in figure 4 were obtained using the stable isotope information by constraining the model parameters during calibration. In a first step, the model is calibrated using only the rainfall-runoff data utilised for model $M_1$ for both catchments. However, the model prediction results were evaluated only in terms of runoff simulation performance. The calibration results show high model performances (see table 7.3). Nevertheless, model performance and identifiability of parameters alone could not assure the adequacy of appropriate model structure (Uhlenbrook et al., 1999; Wagener et al., 2003). It is important to note that the model simulation results using isotope information is conceivable. This is due to the fact that the isotopes provide additional prior information about the proportions of different runoff components in the catchment (Uhlenbrook et al., 2002; McGuire et al., 2005).

For instance in the Chemoga catchment, the model structure $M_1$ (fig. 7.3) was conceptualised with the assumption that rapid sub-surface flow is the dominant runoff generation mechanism in this catchment. Thus, it was hypothesised that rapid sub-surface flow joining the streamflow more or less immediately after the rainfall events. Although, the modelling results were good in terms of performance measures, the isotopes study by Tekleab et al. (2014b) in the same catchment reveal the dominance of overland flow component during the main rainy season. Thus, the stable isotope information provides information about the flow pathways to improve the initial process conceptualization by including the overland flow component. Thus, the isotope data help to refine our earlier perception of the model structure in a way that the parameters are conditioned using percentage fraction of new and old water components and inferred flow pathways in these catchments.

Similarly, our initial assumption in the Jedeb catchment was that the parameter $\alpha$ set to 1 to conceptualise that the slow groundwater flow component was completely fed by the vertical percolation flux $P_{max}$, from the unsaturated zone to the saturated zone and the quick flow component as overland flow was considered to be dominating (100%) during rainfall events. This is because the parameter $P_{max}$, (maximum percolation rate) is obviously important during the long dry season (October to February) and during the spring season (March, to May) to sustain persistent low flow.

Thus, the contribution from the slow flow component during the main rainy season was little to meet the proportion obtained by the use of isotope as depicted by the wet season hydrograph separation. The hydrograph separation results using isotopes exhibit conceivable proportion with a range of values of slow flow (36-48%) and quick flow (52-64%) components during the wet season for the Jedeb catchment. This in turn highlights the weakness of the earlier model structure $M_1$ and demonstrates the importance of the parameter $\alpha$ for the new model structures (see fig. 7.3) represented by $M_2$ model architecture that account fraction of rapid sub-surface flow and $M_3$ further included to capture the larger peaks. Overall, the modelling results using isotope as auxiliary data yielded good results. The results demonstrates the potential of this data similar to previous studies that also advocate the power of stable environmental isotope data in rainfall-runoff modelling (e.g. Uhlenbrook and Hoeg, 2003; Seibert and Mc Donnell, 2002; Son and Sivapalan, 2007; Birkel et al., 2010).

### 7.4.3 The dominant runoff mechanism

In this study, different model representations and stable isotope data were used to better understand the behaviour of the two catchments. It was demonstrated, that the hydrologic systems could not be modelled well by the same model structure due to the differences in the hydrologic processes and behaviour. Consequently, application of stable isotope for wet season hydrograph separation in combination with testing different model structures has provided insights about the dominant runoff generation mechanism in these two catchments. It could be shown that overland flow is the dominating and flood runoff generation mechanism, in particular saturation excess is important after a certain threshold is exceeded which became apparent by the peak flows simulations (fig. 7.8). This suggests that the volume of runoff exceeding the maximum capacity of fast reservoir is modelled as saturation excess overland flow. However, we cannot generalize that the dominant flood runoff generation mechanism in the whole Abay/Upper Blue Nile basin is saturation excess overland flow, as within the basin and even within neighbouring catchments hydrologic behaviour varies considerably (e.g. Kim and Kaluarachchi, 2008; Uhlenbrook et al., 2010; and Tekleab et al., 2011). Hence, more detailed process and research accompanied by the application of more process based models at various temporal and spatial scales would be necessary for detailed understanding of the rainfall-runoff processes in the basin.

The presented results are in line with the studies by Liu et al. (2008b); Collick et al. (2009), and Steenhuis et al. (2009), who have shown that saturation excess over land flow mechanism is the dominant runoff process in most small catchments in the Ethiopian highlands. Though, processes descriptions of the applied models are different from the present modelling approach. For example, Steenhuis et al. (2009), classified the runoff contributing area into three classes viz. hillslopes (precipitation runs off), contributing area rock outcrops, and contributing area saturated valley bottoms. In their approach, in the upland area direct precipitation runoff is conceptualized. In the mid section for the landscape interflow and percolation are hypothesized, and saturation excess over land flow is conceptualized at the lower portion of the landscape. However, it was observed in the field and we argue that flood runoff is generated as a result of saturation excess overland flow even in the upland areas of the mountains or the hillslope close to the river.

The modelling results in a daily time step given in this paper are in agreement with similar hydrological studies in the region in terms of the obtained model performances using Nash-Sutcliffe efficiencies ($E_{NS}$) varies between 0.52-0.84 well comparable with Uhlenbrook et al. (2010) HBV model application in the Gilgel Abay catchment of Lake Tana basin, where $E_{NS}$ varies between 0.6-0.91; Collick et al. (2009) applied relatively simple semi-distributed model on three micro-catchments (catchment area varies between 1.13-4.8 km$^2$) of the Soil Conservation Research Projects within the Abay/Upper Blue Nile basin, $E_{NS}$ varies between 0.56-0.78. Wale et al. (2009) applied the same HBV model in the Lake Tana basin, and the obtained $E_{NS}$ varies between (0.09-0.85). With more complex model SWAT application in the Lake Tana basin, Setegn et al. (2008) obtained $E_{NS}$ varies between 0.04-0.7.

In all of the previous modelling efforts, the performance measure $E_{NS}$ gave well reasonable values in terms of matching the observation with the model. However, as pointed out by Gupta et al. (1998); Schaefli and Gupta (2007) single objective measures $E_{NS}$ is not desirable due to the fact that the information content of the data explaining the different parts of the hydrograph could not be fully explored during calibration process. Hence, it does not tell the actual process control that has been taken place in a catchment.

The main difference between the model structures used in this paper and Steenhuis et al. (2009) ; Tilahun et al. (2013a,b, 2014) is that in the present model, the overland flow is taken in to account through threshold behaviour of the effective rainfall, which primarily satisfied the unsaturated reservoir and the rainfall excess beyond the maximum soil moisture capacity was modelled as fast linear reservoir. Thus, in contrast to the above published papers the field knowledge suggests that shortly after rainfall events, saturation of the agricultural land has occurred in the hillslope due to limited soil moisture capacity of the soil.

In counterpart models, degraded land is assumed to be located in the most upper part of the landscape. This degraded area and the saturated area in the lower portion of the landscape assumed to be the only runoff source areas. But, as far as most of the highland in the study catchments concerned, degraded land is found throughout the landscape in different land use. Hence, the first order streams that generated streamflow can not be modelled well with their assumption that no slow flow component was considered. Furthermore, the interception component and the wetland did not explicitly accounted in the previous models. In general the constitutive relationships linking fluxes, states and stokes are different (see equations 7.3-7.13).

Unlike the previous applied models the approach in the present study takes distributed fluxes, different processes, and additional data of the different runoff components during calibration into account, while remaining a simple conceptual model with less data requirements. Nevertheless, there is a need for further process research supported by tracers, field observation and accurate measurements of precipitation, discharge, groundwater levels, and soil moisture, which are crucial for better understanding of the runoff generation mechanism in all sub-catchments of Abay/Upper Blue Nile basin.

## 7.4.4 Sensitivity and Uncertainty Analysis

The results of the parameter sensitivity analysis in the Chemoga and Jedeb catchments are shown in figures 7.6 and 7.7, respectively. The cumulative normal distribution of parameter sets with the objective function with respect to high and low flows are shown for the different model structures $M_1$ (top), $M_2$ (middle) and $M_3$ (lower). The grey straight line in the figures indicates a uniform distribution when parameters are randomly sampled from the pre-defined feasible parameter space. An insensitive parameter would produce a straight line and the gradient of the cumulative distribution measure the parameter sensitivity (Wagener et al., 2003).

In the Chemoga catchment, the parameters $\beta$, $K_f$, $\alpha$, $S_{w,th}$ depict higher sensitivity. While, parameters $S_{max}$, and $S_{s,th}$ have shown medium sensitivity. The parameters $C$ and $P_{max}$ were least sensitive for both performance measures. The parameter $\beta$ has a key role in transforming rainfall into runoff and controlling the spatial variability of the soil moisture in both catchments. This parameter is identifiable with respect to both objective measures.

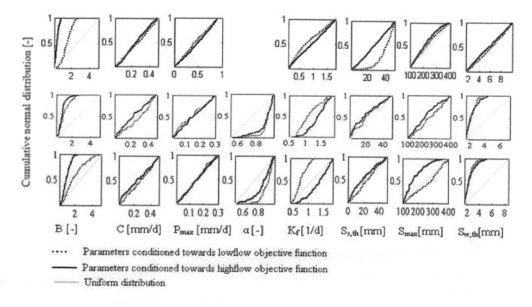

B [-]        C [mm/d]   $P_{max}$ [mm/d]   $\alpha$ [-]      $K_f$ [1/d]   $S_{s,th}$[mm]   $S_{max}$[mm]   $S_{w,th}$[mm]

···· Parameters conditioned towards lowflow objective function
—— Parameters conditioned towards highflow objective function
—— Uniform distribution

Figure 7.6: Sensitivity plots of the model parameters ($M_1$, $M_2$, and $M_3$) sequentially related to high flow objective function represented by the black solid line and low flow objective function represented by the black dot line for the Chemoga catchment.

In figure 7.6, it can be noticed that the degrees of parameter sensitivity was different among $M_1$ versus $M_2$ and $M_3$. Most of the parameters from $M_1$ are insensitive as compared to $M_2$ and $M_3$.An increase in sensitivity of parameters is seen as we move from $M_2$ to $M_3$. This might suggest that the explicit inclusion of the interception component has an effect on the sensitivity results of the other parameters. In general, the parameters $\alpha$, $K_f$, $S_{max}$, and $S_{w,th}$ are better conditioned using isotope information than in the case of $M_1$ using only rainfall-runoff data.

In the Jedeb catchment, the parameters, $K_f$, $S_{f,max}$, and $\alpha$ exhibit higher sensitivity. While the remaining parameters show less sensitivity under high flow conditions. However, the parameter $P_{max}$ reveals relatively higher sensitivity and appears to be the most influential parameter during low flow conditions. The higher sensitivity of $\beta$, $K_f$, $\alpha$, $S_{w,th}$, and $S_{f,max}$ for both performance measures indicate that these parameters are equally important during different response modes of the system in both catchments. Overall the degrees of sensitivity of parameters in the Jedeb catchment are less in $M_1$ and more sensitive in $M_2$ and $M_3$. For example, the high flow parameter $S_{f,max}$, and the low flow parameter $P_{max}$ exhibit more sensitivity in $M_2$ and $M_3$.

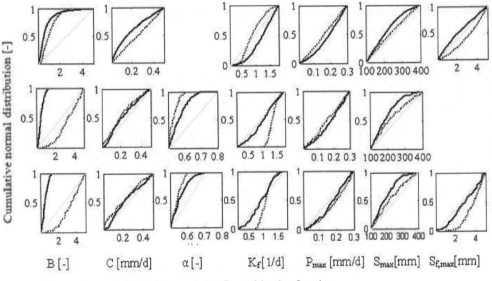

B [-]          C [mm/d]        $\alpha$ [-]          $K_f$[1/d]      $P_{max}$ [mm/d]  $S_{max}$[mm]  $S_{f,max}$[mm]

••••    Parameters conditioned towards lowflow objective function

⎯⎯    Parameters conditioned towards highflow objective function

⎯⎯    Uniform distribution

Figure 7.7: Sensitivity plot of the model parameters ($M_1$, $M_2$, and $M_3$) sequentially related to high flow objective function represented by the black solid line and low flow objective function represented by the blacked dot line for Jedeb catchment.

Figure 7.8 shows the 95% discharge prediction uncertainty results from a MC analysis for the different model representation in both catchments. Among 10,000 MC runs in the Chemoga catchment, for $M_1$ with threshold rejection criteria greater than 0.7, 4707 behavioural parameter sets were used for the discharge prediction uncertainty. For $M_2$ and $M_3$ model representation about 581 and 1296 behavioural sets fulfilled the fraction of new and old water percentage flow. Similarly, in Jedeb catchment for $M_1$ with the same threshold rejection criteria greater than 0.7, 1080 behavioural parameter sets was used for the discharge prediction uncertainty. However, for $M_2$, and $M_3$ model representation in the Jedeb catchment, the number of behavioural parameters sets that met the criteria was found to be 1841 and 1223, respectively.

Noticeable differences in prediction bands among the different model representation are seen in both catchments. Furthermore, the uncertainty band width is decreasing from $M_1$ to $M_3$ in both catchments. This suggests that the predictive uncertainty is decreasing for the model representation $M_3$. In the Jedeb catchment, the uncertainty band is much greater in $M_1$, and most of the peak discharges were not fully captured by $M_2$ and the observed discharge is out of the prediction band. However, the peaks were well captured by $M_3$ and the relative uncertainty band width is decreasing.

To summaries, it is found that the performance measures, the sensitivity, and the predictive uncertainty were different for the different representations $M_1$, $M_2$, and $M_3$ in both catchments. The different model representations using rainfall-runoff data and additional information of stable isotopes (i.e. new and old water contributions) offer the possibility to decide on the better model representation not only based on the model performances but also based on the adequacy of the model structure in terms of both identifiability and reduced predictive uncertainty. Model $M_3$ represented the hydrological system better in both catchments. The results might be further improved with more detailed additional information on groundwater levels, fine resolution isotope data and soil moisture data in further research.

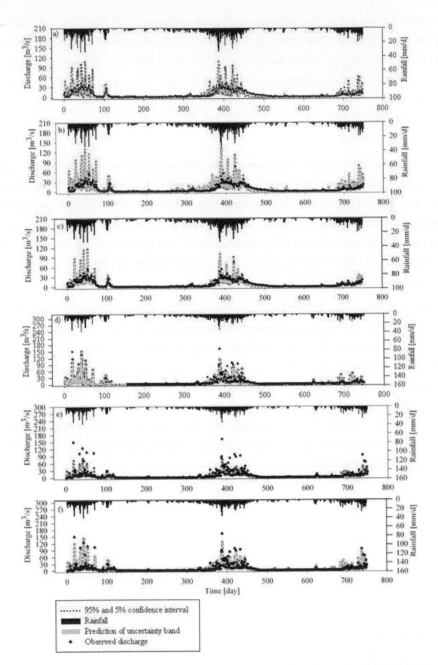

Figure 7.8: Predictive uncertainty plot. Sub plot a, b, and c are for the Chemoga catchment, while d, e, and f are for the Jedeb catchment representing $M_1$, $M_2$, and $M_3$, respectively.

**7.4.5 Limitations of the method**

The main limitation of the approach used is that the parameters are conditioned only with the wet season hydrograph separation results obtained by stable environmental isotope data. Furthermore, different quantitative information such as groundwater levels or soil moisture data is lacking to further increase the trust in the model structure. Another limitation is that during the hydrograph separation, the end member isotope concentrations were prone to uncertainty, which does not give exact values of the different fractions of runoff components (see e.g. Uhlenbrook and Hoeg, 2003). Another shortcoming of the method is the need for more computational time that required to complete sufficient Mote Carlo simulations due to the grid size of the PCRaster model coupled with MCAT in MATLAB.. Thus, the method used has an influence to test different model structures within a short period of time. However, the methodology still gives promising results with regard to retaining behavioural models that deemed to be avoiding the subjectivity of selecting the threshold value of the likelihood measure in the GLUE methodology. The isotope data coupled with the modelling and further field observations provided considerable insights into the different runoff generation processes and runoff components.

## 7.5 Conclusions

The objective of this study was to understand the dominant hydrological processes in two headwater tributaries of Abay/Upper Blue Nile basin. A conceptual distributed modelling approach was used in combination with different field observations. Different model representations were employed to represent the hydrological processes. The application of environmental isotope data have been used to constrain model parameters during calibration by separating the total runoff into fractions of new and old water components during the wet season. The different model representations give different results in terms of the ability to produce the observation, parameter sensitivity, and predictive uncertainty.

Based on the processes description, parameter identifiability, and predictive uncertainty, the $M_3$ model representations seem to be adequate in describing the hydrological behaviour in both catchments. The results demonstrate the dominance of surface runoff generation during events in both catchments. This is in agreement with the field observation that the prevalent effect of soil erosion is due to the dominance of overland flow.

It has been demonstrated that the isotope data is useful to constrain model parameters, which in turn give further insights into the dominant runoff components. The isotope data as an auxiliary information enabled selecting a suitable model structure and reduced the predictive uncertainty. Moreover, the modelling results show that the two catchments cannot be modelled by the same model representation owing to differences in landscape composition and related hydrologic systems behaviour. Therefore, a single model representation for the whole Abay/Upper Blue Nile is not sufficient to capture the dominant hydrological processes in the various sub-catchments. Landscape based modelling, taking account of the proportion of dominant runoff generating mechanisms of different landscape units, may offer a promising venue (Gharari et al, 2013; Gao et al., 2014) in this regard.

Future research will attempt to improve the understanding of the space-time variability of hydrological processes and the modelling is essential. The results will be further improved with additional field observations such as groundwater levels, soil moisture and fine resolution of spatial and temporal variation of environmental isotope data incorporating into modelling process. Furthermore, for better understanding of the hydrological behaviour of the whole basin, similar research in different tributaries of the Abay/Upper Blue Nile basin is needed as well as ways to regionalise small scale process understanding.

# CHAPTER 8

## Conclusions and Recommendations

### 8.1 Conclusions

Understanding hydrological processes and hydro-climatic variability in various space and time dimensions are vital for the management of water resources in the Abay/Upper Blue Nile basin. The water resources in the basin are increasingly important due to the competing demands for agriculture, hydropower, industry, households and environment. Owing to the interdependencies of the water users upstream-downstream, research on hydrology is indispensable for sustainable water resources development, prediction of extremes, and assessment of water quality and quantity. Hydrological studies are limited in the Abay/Upper Blue Nile basin. Previous studies are mainly based on secondary data sets for catchment process understanding and modelling. In this study, a combined use of secondary and primary data sets including experimental studies using stable environmental isotopes enabled to better understand and characterise the hydrology of the basin. Additionally, hydrological modelling using a top-down model approach has an advantage in such data scarce regions due to low data demand. The modelling approach developed in this study resulted in a low parameterized model, well identifiability of parameters, and reduced predictive uncertainty.

This thesis investigated the characterisation and quantification of catchment processes and modelling in the Abay/Upper Blue Nile basin at different temporal and spatial scales, from meso-scale to full basin scale, and from daily to annual time step. To assess, long-term trends of main forcing parameters on runoff, statistical analyses have been conducted for rainfall, temperature and runoff. How land use/land cover change has impacted long-term trends of runoff was assessed by a conceptual hydrological model. A Budyko type model has been used to understand the water balance relationships from meso-scale up to basin scale. To complete the picture at local scale, detailed hydro-meteorological measurements, including isotope data have been collected for three years in two headwater meso-scale catchments (Jedeb and Chemoga). These data helped to develop detailed conceptual distributed models to understand better the rainfall-runoff processes in these two catchments using PCRaster software environment.

This study investigated the long-term trends of hydro-climatic variables in the Abay/Upper Blue Nile basin (Chap.3). The Mann-Kendall and Pettitt tests have been used in the analysis. The results of trends of hydrological variable in the basin depict a number of significant trends, both increasing and decreasing. With regard to discharge time series, the study could not find any consistent change or patterns of trends among the investigated catchments. Thus, the results of the trends across the investigated catchments are heterogeneous. For instance,

neighbouring catchments, Gilgel Abay and Koga as well as Gumera and Rib in the Lake Tana
sub basin showed contrasting trends and change points in the observed series. The difference
in the results of the trend and change point directions might be attributed to the degree of
human interventions and catchment physiographic characteristics (land use/land cover, soil
type, topography and geology) in different catchments.

The investigation of precipitation in mean annual or seasonal time scales did not show
statistically significant discernible trends evaluated at 5% significance level across the
examined precipitation stations within the basin. However, the minimum, maximum and
average temperature in various stations in the basin reveals statistically significant increasing
trends. For instance, increasing trends in temperature at different weather stations for the
mean annual, rainy, dry and short rainy seasons are apparent. Therefore, it is concluded that
trends and change point times varied considerably across the examined stations. Moreover,
the identified significant trends can help to make better planning decisions for water
management. However, to attribute the causes to the observed changes in hydro-
meteorological variables require further research (e.g. land use and climate changes). Hence,
good quality of hydro-meteorological data is needed for better management of water
resources in the region.

Furthermore, this study also investigated the water balance dynamics of twenty catchments
based on the Budyko hypothesis (Chap. 4). The model was implemented at spatial scales of
meso-scale to large scale and at annual and monthly temporal scales. Following a top-down
modelling approach, the annual water balances of twenty catchments have been analysed
based on the climate control and atmospheric demand with the assumption that the simple
aridity index annual potential evaporation to annual precipitation ($E_p/P$) is able to explain the
annual water balance.

The simulation results of the model at annual time scale did not reproduce the observed
annual runoff and actual evaporation well. The twenty catchments are scattered with distinct
evaporation ratios (E/P) (see fig 4.4). The different evaporation ratios explain the differences
of the catchments in partitioning the available water into annual runoff and evaporation. The
annual runoff ratio of the examined catchments varies from 0.21-0.7 implying that catchments
have different response patterns for the rainfall inputs. Different reasons are expected to
explain scatter of the catchment represented by different curves. The rainfall seasonality,
inter-annual rainfall variability, non-stationary conditions in the catchments e.g. land use/land
cover changes could be among the reasons for the scatter. Based on the aridity index, most of
the investigated catchments in the basin are towards an intermediate environmental condition
(i.e. neither energy limited nor moisture constrained) in which the aridity index varies
between 0.57-1.5.

The model simulation results at annual time scale are improved at a monthly time scale by inclusion of the soil moisture dynamics into the model. The model simulation at monthly time scale is capable of capturing the monthly peak and low flows reasonably well in the majority of the catchments during calibration and validation periods. Moreover, the majority of the model parameters are well identifiable while explaining the monthly flow dynamics.

The study concluded that, the water balance model results have practical importance, due to the fact that in data scarce regions lower parameterized models are advocated as they require little input data. Thus, owing to parsimonious parameterization, the application of such a water balance model can be very importance as it can support water resources managers in assessing catchment water resources.

The model based on the Budyko approach in conjunction with statistical tests was further used to study the effects of land use/land cover change on hydrologic responses of the meso-scale Jedeb catchment (Chap. 5). The results from the statistical analysis depict a number of statistically significant trends detected in flow variability indicators. These statistically significant changes in daily flow variability indicators reveal significant changes in the flow hydrograph. Peak flow is enhanced, i.e. the response appears to be flashier. There is a significant increase in the rise and fall rates of the flow hydrograph, as well as the number of low flow pulses below a threshold level. The discharge pulses show a declining duration with time. The model results also showed that the soil moisture retention parameter is decreasing over time. The results in general provide useful scientific evidences about modified runoff hydrograph in the headwater catchment. These combined results suggest that land use management practices play a more considerable role on changes in the (daily) flow hydrograph than the climate (i.e. rainfall).

Therefore, it can be concluded that mismanagement of agricultural practices in unfavourable land, and limited adoption of soil conservation measures are the major factors for changes in the flow hydrograph. Adopting appropriate farming practices (e.g. ploughing along the contour, deep ploughing), and land use management (e.g. implementing soil conservation structures, crop patterns and varieties) could help reducing surface runoff generation by augmenting groundwater recharge. This can be a practical mitigation measure for sustainable use of land and water in this catchment. Similar studies in the remaining sub-catchments in the basin would be crucial from the perspective of sustainable water resources management in the region. In this regard, to gain insights about the detailed process knowledge, the experimental research supported by stable environmental isotopes offers a possibility to identify flow pathways and the dominant runoff components in the headwater catchments.

Hence, in this study, characterisation of stable environmental isotopes has been done to identify mean residence times and contributions of runoff components in the meso-scale, agricultural dominated Chemoga and Jedeb catchments (see Chap. 6). Two component mixing model on a seasonal timescale and a sine wave regression approach were used for identifying and quantifying runoff components and mean residence times.

Applications of the isotope method did provide useful information about the potential moisture source areas, and insights about the runoff components in the studied catchments. The results of the analysis showed that the isotopic composition of precipitation exhibit marked seasonal variations, which show distinct isotope signature in three seasons. This suggests different sources of moisture generation for the rainfall in the study area. The results of the two component mixing model demonstrated the dominance of overland flow for streamflow generation in the studied catchments. Besides, the results from the sine wave regression approach showed that both catchments have short mean residence times of 4.1 and 6 months in the Chemoga and Jedeb, respectively.

It is concluded that, the application of stable environmental isotopes in the study area has both scientific and practical significance in terms of water resources management and planning. For instance, the estimation of the mean residence times in the order of months suggests a young age of the shallow groundwater system (fast turnover time). Furthermore, the dominance of overland flow components coupled with agricultural farming practices in steep unfavourable terrrain led to soil erosion, sedimentation, and worsening of the water quality in these meso-scale catchments. This is likely similar in other meso-scale catchments in the basin. Therefore, this study has significant relevance for the region to better inform water management and planning. Furthermore, the study can be used as a base line to extend similar hydrological studies supported by isotopes in the basin. Additionally, the results from the isotope study are useful in providing information about flow pathways and contribution of different runoff components for better understanding of the rainfall-runoff processes.

Finally, rainfall-runoff modelling on daily timescale was carried out in the two meso-scale agricultural dominated catchments Chemoga and Jedeb (see Chap.7). A conceptual distributed model based on box model approaches (e.g., Fenicia et al. 2006, 2008a and 2008b) was developed to study the rainfall-runoff processes and identifying the dominant runoff generation mechanism. Furthermore, different conceptual distributed model representations were implemented in the PCRaster software modelling environment. The models were evaluated based on the performance, i.e. the ability to reproduce the observations, parameter identifiability and predictive uncertainty.

In a first step, the model parameters were calibrated using only rainfall and runoff data. Then, as an additional constraint, the results of seasonal hydrograph separation using isotopes (percentages of runoff components) have been used to calibrate the model parameters (Tekleab et al., 2014b). The model simulation results based on the $M_3$ representation (see Figure 7.3 and 7.5) seems to be adequate in describing the hydrological processes of both catchments. Moreover, the modelling results supported by field observations demonstrated that saturation excess overland flow is the dominant runoff generation mechanism in these catchments during events. Nevertheless, the two catchments cannot be modelled by the same model representation owing to differences in landscape composition and related hydrologic systems behaviour. Therefore, a single model representation for the whole Abay/Upper Blue Nile is not suitable to capture the dominant hydrological processes in the various sub-catchments. Landscape based semi-distributed modelling (applying the FLEX-Topo approach) taking into account the individual runoff processes assigned into different landscape units seem to offer a promising avenue for process-based modelling in the study area (Savenije, 2010; Gharari et al, 2013; Gao et al., 2014).

Overall, combining primary data of climate, water levels, stable isotopes, field observations, and secondary data in modelling is an appropriate approach for the understanding the dynamics and management of limited water resources in the Abay/Upper Blue Nile basin and likely in the remaining basins in the Nile region.

## 8.2 Recommendations

This thesis characterised and quantified catchment processes and runoff components and advanced process-based modelling in the Abay/Upper Blue Nile basin at different temporal and spatial scales, from meso-scale to full basin scale and from daily to annual time step. Thus, the thesis provided in-depth understanding of the basin hydrology supported by new measurements of precipitation, water levels and experimental studies using stable environmental isotope data.

The water balance results based on Budyko modelling showed heterogeneous responses in the Abay/Upper Blue Nile basin, and detailed measurements and modelling in two meso-scale catchments have been done to gain futher insights into process behaviour. However, this is not enough to fully understand the spatial and temporal variability of the dominant hydrological processes and runoff generation mechanism in the basin. On the one hand, the heterogeneities in climate, topography, soil, vegetation and geology could influence the responses of the catchments. On the other hand, change of processes at temporal and spatial scales prohibits simple generalization about the dominant runoff generation mechanism in the basin (Beven, 2000). Consequently, to gain further insights about the runoff generation mechanisms in the remaining sub-catchments, more in-depth studies on runoff generation processes are recommended. Hence, intensive data collection (e.g. fine resolution of water sampling for isotope analysis during events, precipitation, and water level for streamflow discharge, groundwater levels, soil moisture, and remotely sensed data) are indispensable. Moreover, testing the model described in (Chap.7) using PCRaster modelling environment on the experimental Soil Conservation Research Project (SCRP) study sites (e.g. Anjeni, Mybar, Adit Tid and Yeku) within the basin would be crucial for better understanding of the hydrological processes at micro catchment scale. Therefore, further modelling studies are recommended to up-scale the processes understanding from the small scale to the catchment and basin scale. This is essential for better understanding of the hydrological system dynamics in the basin and prediction of environmental changes in the future. Moreover, intensive soil conservation measures aiming at reducing soil erosion and appropriate land use zoning or demarcation (e.g. agricultural area, grazing land, bare land, and forest) is recommended for sustainable land water management in the basin.

Overall, the data scarcity problem in combination with its poor quality will be mitigated using new measurement techniques in combination with comprehensive data collection and improved spatio-temporal coverage. Furthermore, continuous data quality monitoring (e.g. regularly updating rating curves at the discharge gauges during different flow regimes, skilled and responsible human resources, who meticulously operating both the field and office work, and inter-stations comparison of rainfall measurements on different seasons) would be essential to reduce the input data uncertainty and enhance further hydrological research and water mangement in the region.

In this regard, the Ethiopian Ministry of Water and Energy and the National Meteorological Agency should work together on the data collection as well as measures related to quality assurance, archiving and data protocol to be used for research and development. Progress in the data situation could be achieved through collaborative efforts made by the pertinent governmental organizations in particular the Ethiopian Ministry of Water and Energy, National Meteorological Agency, the regional water resources offices, international stakeholders (e.g. ENTRO, NBI) and the rural communities at large.

# REFERENCES

Abbott, M., Bathurst, J., Cunge J.A., O'Connell, P., and Rasmussen, J. (1986). An introduction to the European Hydrologic system (SHE), Structure of a physically based distributed modelling system. Journal of Hydrology, 87: 61-77.

Abdo, K.S., Fisha, B.M., Rientjes, T.H.M., and Gieske, A.S.M. (2009). Assessment of climate change impacts on the hydrology of Gilgel Abay catchment in Lake Tana basin, Ethiopia Hydrol. Process., 23, 3661–3669, doi: 10.1002/hyp.7363.

Abdual Aziz, O.I and Burn, D.H. (2006). Trends and variability in the hydrological regime of the Mackenzie River Basin, Journal of hydrology, 319, 282–294.

Abtew, W., Melesse, A.M., and Dessalegne, T. (2009). Spatial, inter and intra-annual variability of the Upper Blue Nile Basin rainfall. Hydrol. Process, 23, 3075–3082, doi: 10.1002/hyp.7419.

Ali, G., Oswald, C.J., Spence, C., Cammeraat, E.L.H., McGuire, K., Meixner, T., and Reaney, S.M. (2013). Towards a unified threshold-based hydrological theory: necessary components and recurring challenges. Hydrological Processes, 27, 313–318.

Allen, R. G., Pereira, L. S., Raes, D., and Smith, M. (1998). Crop Evapotranpiration: Guild lines for computing crop water requirements, FAO Irrigation and Drainage Paper No 56. Food and Agriculture Organization, Land and Water. Rome, Italy.

Archer, D.R. (2000). Indices of flow variability and their use in identifying the impact of land use changes. In: Proceedings of 7th National Hydrology Symposium, British Hydrological Society, Sept. British Hydrological Society, London, pp. 2.67–2.74.

Archer, D.R. (2007). The use of flow variability analysis to assess the impact of land use change analysis on the paired plynlimon catchments, mid-Wales. Journal of hydrology, 347 487-496.

Aravena, R., Suzuki, O., Pena, H., Grilli, A., Pollastri, A., and Fuenzalida, H. (1999). Isotopic composition and origin of the precipitation in Northern Chile, Appl. Geochem., 14:411–422.

Arsano, Y., and Tamirat, I. (2005). Ethiopia and the Eastern Nile basin. Aquat. Sci., 67 (2005) 15–27, 1015-1621/05/010015-13, doi: 10.1007/s00027-004-0766-x.

Atkinson, S., Woods, R.A., and Sivapalan, M. (2002). Climate and landscape controls on water balance model complexity over changing time scales. Water Resources Research, 38(12), 1314, doi: 10.1029/2002 WR 001487, 50.1-50.17.

Baker, B.D., Richards, R.P., Loftus, T.T., and Kramer, J.W. (2004). A new flashiness index: characteristics and applications to Midwestern rivers and streams. Journal of the American Water Resources Association (JAWRA) 40(2):503-522.

Barthold, F.K., Wu, J., Vache, K.B., Schneider, K., Frede, H.G., and Breuer, L. (2010). Identification of geographic runoff sources in a data sparse region: hydrological processes and the limitations of tracer-based approaches. Hydrol. Process, 24, 2313-2327, doi: 10.1002/hyp.7678.

Bastiaanssen, W.G.M., Menenti, M., Feddes, R.A., and Holtslag A.A.M. (1998). A remote sensing surface energy balance algorithm for land (SEBAL). Part 1. Formulation. Journal of hydrology, 212/213, 198-212.

BCEOM. (1998a). Abay River Basin Integrated Development Master Plan-Phase 2 – Data collection Site investigation survey and analysis, section II Volume I (part 1 Geology), Ministry of Water Resources, Addis Ababa, 103 pp.

BCEOM. (1998b). Abay River Basin Integrated Development Master Plan-Phase 2-Land Resources Development–Reconnaissance Soils Survey, Ministry of Water Resources, Addis Ababa, 208 pp.

BCEOM. (1999a). Abay Basin Integrated master plan study, main report Ministry of water resources, Addis Ababa. Phase two, Volume three Agriculture, 1-2.

BCEOM. (1999b). Abay River Basin Integrated Development Master Plan-Phase 3 – Main report, Volume I, Ministry of Water Resources, Addis Ababa, 435 pp.

BergstrÖm, S. (1995). The HBV model. In V. P. Singh (Ed.), Computer models of watershed hydrology (p. 443-476). Water Resources Publications, Highlands Ranch, CO.

Betrie, G.D., Mohammed, Y. A., Van Griensven, A., and Srinivasan. R. (2011). Sediment management modelling in the Blue Nile Basin using SWAT model. Hydrol. Earth Syst. Sci., 15, 807–818, doi: 10.5194/hess-15-807.

Beven, K. and Kirkby, M. (1979). A physically based, variable contributing area model of basin hydrology. Hydrological Sciences Bulletin, 24(1): 43-69.

Beven, K., and Binley, A. (1992). The future of distributed models: Model calibration and uncertainty prediction. Hydrological Processes, 10,769 -782, 2006 Vol. 6, 279-298.

Beven, K, (2000). Uniqueness of place and process representations in hydrological modelling. Hydrology and Earth System Sciences, 4: 203–213.

Beven, K. J. (2001). Rainfall-runoff modelling: the primer, John Wiley & Sons, Ltd., Chichester, West Sussex, England, U.K.

Beven, K.J. (2002a). Toward an alternative blueprint for a physically based simulated digitally hydrologic response modelling system. Hydrological processes hydrol. process. 16 189-206, doi: 10.1002/hyp.343.

Beven, K.J. (2002b). Towards a coherent philosophy for modelling the environment. Proceedings of the royal society of London. Series A, Mathematical and Physical sciences, 458, 1-20.

Beven, K, (2008). On doing better hydrological science, Hydrological Processes, 22: 3549–3553.

Bewket, W., and Sterk, G. (2005). Dynamics land cover and its effect on the stream flow on the Chemoga watershed in the Blue Nile basin, Ethiopia. Hydrol. Process., 19, 445-458.

Birkel, C., Tetzlaff, D., Dunn, S.M., and Soulsby, C. (2010). Towards a simple dynamic process conceptualization in rainfall-runoff models using multi-criteria calibration and tracers in temperate, upland catchments. Hydro. Process. 24, 260-275, doi: 10.1002/hyp.7478.

Blöschl, G., and Sivapalan, M. (1995) Scale issues in hydrological modelling: A review. Hydrological Processes 9 (3-4): 251-290.

Birsan, M.V., Molnar, P., Burlando, P., and Pfaundler, M. (2005). Streamflow trends in Switzerland. Journal of Hydrology, 314, 312-329.

Bliss, C.I. (1970). Periodic Regressions Statistics in Biology. McGraw-Hill Book Co., New York, USA, pp. 219–287.

Blöschl, G., and Sivapalan, M. (1995). Scale issues in hydrological modelling: A review. Hydrological Processes 9 (3-4): 251-290.

Blöschl, G. and Zehe, E. (2005). On hydrological predictability. Hydrological Processes, 19, 3923–3929.

Bosch, J.M., and Hewlett, D. (1982). A review of catchment experiments to determine the effects of vegetation changes on water yield and evapotraspiration. Journal of Hydrology, 170: 123–135.

Bowden, G.J., Dandy, G.C., and Maier, H.R. (2005). Input determination for neural network models in water resources applications. Part 1-background and methodology. J. Hydrol. 301, 93-107.

Boyle, D. (2000). Multi-criteria calibration of hydrological models (PhD thesis). Dept. of Hydrology and Water Resources, University of Arizona, Tucson.

Boyle, D.P., Gupta, H.V., and Sorooshian, S. (2000). Toward improved calibration of hydrologic models: Combining the strengths of manual and automatic methods, Water Resour. Res., 36, 3663–3674.

Botter, G., Bertuzzo, E., and Rinaldo, A. (2011). Catchment residence and travel time distributions: The master equation. Geophys. Res. Lett., 38, L11403, doi: 10.1029/2011GL047666.

Bronstert, A., and Plate, E. (1997). Modelling of runoff generation and Soil Moisture Dynamics for hill slopes and Micro-catchments. Journal of Hydrology 198; 177-195.

Brown, A.E., Zhang, L., McMahon, T.A., Western, A.W., and Vertesse, R.A. (2005). A review of paired catchment studies for determining changes in water yield resulting from alterations in vegetation. Journal of Hydrology 310: 28-61.

Budyko, M.I. (1974). Climate and life. Academic, New York.

Burnash, R.J.C., Ferral, R.L. and McGuire, R.A. (1973). A generalized streamflow simulation system, conceptual modelling for digital computers. Report by the Joint Federal State River Forecasting Centre, Sacramento, CA, USA.

Buttle, J.M. (1994). Isotope hydrograph separations and rapid delivery of pre-event water from drainage basins, Progress in Physical Geography, 18: 16–41.

Buttle, J.M., Creed, I.F., and Moore, R.D. (2005). Advances in Canadian forest hydrology, 1999–2003. Hydrological Processes, 19, 169–200.

Buytaert, W., and Beven, K. (2011). Models as multiple working hypotheses: hydrological simulation of tropical alpine wetlands, Hydrol. Process. 25, 1784–1799, doi: 10.1002/hyp.7936.

Camberlin, P. (1995). June–September rainfall in north-eastern Africa and atmospheric signals over the tropics: A zonal perspective. Int. J.Climatol., 15, 773–783.

Camberlin, P. (1997). Rainfall anomalies in the Source Region of the Nile and their connection with the Indian Summer Monsoon. J. Climate, 1380-1392.

Clark, M.P., Slater, A.G., Rupp, D, E., Woods, R.A., Vrugt, H.V., Wagener, T., and Hay, L.E. (2008). Framework for Understanding Structural Errors (FUSE): A modular framework to diagnose differences between hydrological models, Water Resour Res, W00B02, doi: 10.1029/2007WR006735.

Clark, M.P., McMillan, H.K., Collins, D.B.G., Kavetski, D., and Woods, R.A. (2011a). Hydrological field data from a modeller's perspective: Part 2: process-based evaluation of model hypotheses, Hydrol. Process., 25, 523–543, doi: 10.1002/hyp.7902.

Clark, M.P., Kavetski, D., and Fenicia, F. (2011b). Pursuing the method of multiple working hypotheses for hydrological modelling. Water Resour. Res., 47, W09301, doi: 10.1029/2010WR009827.

Claver, A., and Wood, W. (1995). The institute of hydrology distributed model. In computer models of watershed hydrology. Singh VP (ed) Water Resources Publications Colorado; 595-626.

Collick, A.S., Easton, Z.M., Ashagrie, T., Biruk, B., Tilahun, S., Adgo, E., Awlachew, S.B., Zeleke, G., and Steenhuis, T.S. (2009). A simple semi-distributed water balance model for the Ethiopian highlands. Hydrol.process.23, 3718-3727, doi 10.1002/hyp-7517.

Conway, D., and Hulme, M. (1993). Recent fluctuations in precipitation and runoff over the Nile sub-basins and their impact on main Nile discharge. Climatic Change, 25, 127–151.

Conway, D. (1997). Water balance model of the Upper Blue Nile in Ethiopia. Hydrological Sciences Journal 42, pp. 265-286.

Conway, D. (2000). Climate and Hydrology of the upper Blue Nile River Basin. The Geographical Journal volume, 166.No.1, PP. 49-62.

Conway, D., Mould, C., Bewket, W. (2004). Over one century of rainfall and Temperature observation in Addis Ababa, Ethiopia. Int. J. Climatol.24: 77-91.

Conway, D. (2005). From head water tributaries to international river: Observing and adapting to climate variability and change in the Nile Basin. Global Environ. Change, 15, 99-114.

Corzo, G., and Solomatine, D. (2007). Baseflow separation techniques for modular artificial neural network modelling in flow forecasting. Hydrological Science Journal, 52(3).

Costa, M.H., Notta, A., and Cardille, J.A. (2003). Effects of large-scale changes in land cover on the discharge of the Tocantins River, Southeastern Amazonia. Journal of Hydrology 283: 206–217.

Dansgaard, W. (1964). Stable isotopes in precipitation, Tellus, 16, 436–468.

Darling, W.G., and Gizaw, B. (2002). Rainfall–groundwater isotopic relationships in eastern Africa: the Addis Ababa anomaly. Study of environmental change using isotopic techniques. C&S Papers Series, IAEA, pp 489–490.

DeFries, R., and Eshleman, K.N. (2004). Land use change and hydrologic processes: a major focus for the future. Hydrological processes 18: 2183-2186.

de Groen, M.M., and Savenije, H.H.G, (2006). A monthly interception equation based on the statistical characteristics of daily rainfall, Water Resour. Res., 42, W12417, doi: 10.1029/2006WR005013.

Di Baldassarre, G., Elshamy, M., Griensven, A.V., Soliman, E., Kigobe, M., Ndomba, P., Joseph, Mutemi., Mutua, F., Moges, S., Xuan Y., Solomatine, D., and Uhlenbrook, S. (2011). Future hydrology and climate in the river Nile basin: a review. Hydrological sciences journal, 56(2). Doi: 10.1080/02626667.2011.557378.

Didszun, J., and Uhlenbrook, S. (2008). Scaling of dominant runoff generation processes: Nested catchments approach using multiple tracers, Water Resour. Res., Vol. 44, W02410, doi: 10.1029/2006WR005242.

Dincer, T., Payne, B.R., Florkowski, T., Martinee, J, and Tongiorgi, E. (1970). Snow-melt runoff from measurements of tritium and oxygen.18. Water Resour. Res., 6:110-124.

Donohue, R.J., Roderick, M.L., and McVicar, T.R. (2007). On the importance of vegetation dynamics in Budyko's hydrological model. Hydrol. Earth. Syst. Sci., 11, 983-995.

Dunn, S.M., McDonnell, J.J., and Vache, K.B. (2007). Factors influencing the residence time of catchment waters: A virtual experiment approach, Water Resour. Res., 43, W06408, doi: 10.1029/2006WR005393.

Easton, Z.M., Fuka, D.R., White, E.D., Collick, A.S., Ashagre, B.B., Mc Cartney, M., Awlachew, S.B., Ahmed, A.A and Steenhuis, T.S. (2010). A multi basin SWAT model analysis of ruoff and sedimentation in the Blue Nile, Ethiopia, Hydrol. Earth Syst. Sci., 14, 1827–1841, doi: 10.5194/hess-14-1827.

EEPCO, Ethiopian Electric Power Corporation. (2010). 5000 hydroelectric project basic design, Vol. I, main report, 216pp.

Elshamy, M.E., Seierstad, I.A., and Sorteberg, A. (2009). Impacts of climate change on Blue Nile flows using bias-corrected GCM scenarios. Hydrol. Earth Syst. Sci., 13, 551–565.

Eltahir, Elfatih, A.B. (1996). El Ni~no and the natural variability in the flow of the Nile River. Water Resources Research, Vol. 32, N0. 1, pages 131-137.

ENTRO. (2007). Cooperative regional assessment for watershed management. Transboundary Analysis Abay-Blue Nile sub-Basin, Addis Ababa, Ethiopia.

FangFang, Z., Zhang, L., and Xu, Z. (2009). Effects of vegetation cover change on stream flow at a range of spatial scales.18th World IMACS /MODSIM Congress cairns Australia 13-17.

FDREMW, Federal Democratic Republic of Ethiopia, Ministry of Water Resources. (2002). Water sector development programme. Addis Ababa.

Fenicia, F., Savenije, H. H. G., Matgen, P., and Pfister, L. (2006), Is the groundwater reservoir linear? Learning from data in hydrological modelling, Hydrol. Earth Syst. Sci., 10, 139–150, doi: 10.5194/hess-10-139.

Fenicia, F., Savenije, H.H.G., Matgen, P., and Pfister, L. (2007). A comparison of alternative multi objective calibration strategies for hydrological modelling. Water Res, 43, 3434, doi: 10.1029/2006 WR005098.

Fenicia, F., Savenije H.H.G, Matgen, P., and Pfister, L. (2008a). Understanding catchment behaviour through stepwise model concept improvement, Water Resour. Res., 44, W01402, doi: 10.1029/2006WR005563.

Fenicia, F., Savenije, H.H.G., and McDonnell, J.J. (2008b). Learning from model improvement: On the contribution of complementary information to process understanding. Water Resources Research, 44, W06419, doi: 10.1029/2007WR006386.

Freer J, Beven K.J., and Ambroise, B. (1996). Bayesian estimation of uncertainty in runoff prediction and the value of data: an application of the GLUE approach. Water Resources Research 32: 2161–2173.

Fu, B.P. (1981). On the calculation of the evaporation from land surface (in Chinese) Sci. Atmos. Sin., 5(1), 23-31.

Gao, H., Hrachowitz, M., Fenicia, F., Gharari, S. & Savenije, H. H. G. (2014). Testing the realism of a topography driven model (FLEX-Topo) in the nested catchments of the Upper Heihe, China, Hydrol. Earth Syst. Sci. 18, 1895-1915, doi: 10.5194/hess-18-1895.

Gat, J. R., Bowser, C. J., and Kendall, C. (1994). The contribution of evaporation from the Great Lakes to the continental atmosphere: Estimate based on stable isotope data: Geophysical Research Letters, v.21, p.557-560.

Gat, J.R. (1996). Oxygen and Hydrogen isotopes in the hydrologic cycle, Annual Review Earth Planetary Sciences., 24, 225–262.

Gebrehiwot, S. G., Taye, A., and Bishop, K. (2010). Forest cover and stream flow in a headwater of the Blue Nile: complementing observational data analysis with community perception, AMBIO, 39, 284–294, doi: 10.1007/s13280-010-0047-y.

Gebremichael, T.G., Mohamed, Y.A., Betrie, G.D., van der Zaag, P., and Teferi, E. (2012) .Trend Analysis of Runoff and Sediment Fluxes in the Upper Blue Nile Basin: A Combined Analysis of Statistical Tests, Physically-based Models and Land use Maps, Journal of Hydrology, doi: http://dx.doi.org/10.1016/j.jhydrol.2012.12.023.

Genereux, D.P. (1998). Quantifying uncertainty in tracer based hydrograph separation. Water Resources Research. 34 (4): 915-919.

Gerrits, A.M.J., Savenije, H.H.G., Veling, E.J.M., and Pfister, and L. (2009). Analytical derivation of the Budyko Curve based on rainfall characteristics and a simple evaporation model. Water Resources Research, Vol 45 W04403, doi; 10.1029 /2008WR007308.

Gharari, S., Hrachowitz, M., Fenicia, F., and Savenije, H.H.G. (2011). Hydrological landscape classification: investigating the performance of HAND based landscape classifications in a central European meso-scale catchment. Hydrology and Earth System Sciences, 15, 3275–3291, doi: 10.5194/hess-15-3275.

Gharari, S., Hrachowitz, M., Fenicia, F., Gao, H. & Savenije, H. H. G. (2013) Using expert knowledge to increase realism in environmental system models can dramatically reduce the need for calibration, Hydrol. Earth Syst. Sci. Discuss., 10, 14801-14855.

Gibson, J.J., Edwards, T.W.D., Birks, S.J., Amour, N.A.St., Buhay, W.M., McEachern, P., Wolfe, B.B., and Peters, D.L. (2005). Progress in isotope tracer hydrology in Canada. Hydrol. Process.19, 303– 327, doi: 10.1002/hyp.5766.

Gupta, H.V., Sorooshian, S., and Yapo, P.O. (1998). Toward improved calibration of hydrologic models: Multiple and non-commensurable measures of information. Water Resources Research, 34(4), 751-764, 1998.

Haile, A.T., Rientjes, T.H., Gieske, M., and GebreMichael, M. (2009). Rainfall variability over mountainous and adjacent Lake areas: the case of Lake Tana basin at the source of the Blue Nile River. J. Appl. Meteor. Climatol., 48, 1696-1717.

Haile, A.T. (2010). Rainfall variability and estimation for hydrologic modelling. A remote sensing based study at the source basin of the upper Blue Nile River. PhD. thesis. ITC Enschede, ISBN 978 90-6164-286-8. 229pp.

Hargreaves, G.H., and Samani, Z.A. (1982). Estimating potential evaporation. J. Irrig. Drainage Eng., 108(3), 225-230.

Hargreaves, G. H., and Allen, R. G. (2003). History and Evaluation of Hargreaves Evapotranspiration Equation. Journal of Irrigation and Drainage Engineering, 129 (1), 53-63.

Helsel, D.R., and Hirsch, R.M. (1992). Statistical Methods in Water Resources. Studies in Environmental Science 49. Elsevier, Amsterdam, The Netherlands.

Heidbüchel, I., Troch, P.A., Lyon, S.W., and Weiler, M. (2012). The master transit time distribution of variable flow systems, Water Resour. Res., 48, W06520, doi: 10.1029/2011WR011293, 2012.

Hess, A., Iyer, H., and Malm, W. (2001). Linear trend analysis: a comparison of methods. Atmos. Environ. 35, 5211–5222.

Hrachowitz, M., Soulsby C., Tetzlaff, D., Dawson J.J.C., Dunn, S.M., and Malcolm, I.A. (2009). Using longer-term data sets to understand transit times in contrasting headwater catchments, Journal of Hydrology, doi:10.1016/j.jhydrol.2009.01.001.

Hrachowitz, M., Bohte, R., Mul, M. L., Bogaard, T. A., Savenije, H. H. G., and Uhlenbrook, S. (2011a). On the value of combined event runoff and tracer analysis to improve understanding of catchment functioning in a data-scarce semi-arid area, Hydrol. Earth Syst. Sci., 15, 2007–2024, doi: 10.5194/hess-15-2007.

Hrachowitz, M., Soulsby, C., Tetzlaff, I., and Malcolm, A. (2011b). Sensitivity of mean transit time estimates to model conditioning and data availability, Hydrol. Processes., 25, 980–990.

Hrachowitz, M., Savenije, H., Bogaard, T.A., Tetzlaff, D., and Soulsby, C. (2013a). What can flux tracking teach us about water age distribution patterns and their temporal dynamics? Hydrol. Earth Syst. Sci., 17, 533-564, doi: 10.5194/hess-17-533.

Hrachowitz, M., Savenije, H.H.G., Blöschl, G., McDonnell, J.J., Sivapalan, M., Pomeroy, J.W., Arheimer, B., Blume, T., Clark, M.P., Ehret, U., Fenicia, F., Freer, J.E., Gelfan, A., Gupta, H.V., Hughes, D.A., Hut, R.W., Montanari, A., Pande, S., Tetzlaff, D., Troch, P.A., Uhlenbrook, S., Wagener, T., Winsemius, H.C., Woods, R.A., Zehe, E., and Cudennec, C. (2013b). A decade of Predictions in Ungauged Basins (PUB)—a review. Hydrological Sciences Journal, 58(6): 1–58.

Hu, Y., Maskey, S., Uhlenbrook, S, and Zhao, H. (2011). Streamflow trends and climate linkages, in the source region of the Yellow river, China. Hydrological processes, 25, 3399-3411.

Hurni, H. (1988). Principles of soil conservation for cultivated land. Soil Technology, 1: 101-116.

Hurni, H. (1993). Land degradation, famines and resource scenarios in Ethiopia. In World Soil Erosion and Conservation, ed. D. Pimental, pp. 27-62. Cambridge University Press, Cambridge.

Hurni, H., Tato, k., and Zeleke, G. (2005). The implication of changes in population, Land use and land management for surface runoff in the upper Nile basin area of Ethiopia. Mountain Research and Development, Vol 25. No.2 May 2005: 147-154.

International Atomic Energy Agency. (2009). IAEA-WMO Programme on Isotopic Composition of Precipitation: Global Network of Isotopes in Precipitation (GNIP) Technical procedure for sampling.

Jiang, S., Ren, L., Yomg, B., Singh, V.P., Yang, X., and Yuan, F. (2011). Quantifying the effects of climate variability and human activities on runoff from the Laohahe basin in northern China using three different methods. Hydrological Processes. doi: 10.1002/hyp.8002.

Johnson, P.A., and Curtis, P.D. (1994). Water Balance of Blue Nile River Basin in Ethiopia, Journal of Irrigation and Drainage Engineering, Vol. 120, No. 3, 573-590.

Joseph, A., Frangi, P., and Aranyossy, J.F. (1992). Isotopic composition of Meteoric water and groundwater in the Sahelo-Sudanese Zone, Journal of Geophysical research., 97: 7543-7551.

Jothityangkoon, C., Sivapalan, M., and Farmer, D.L. (2001). Process control of water balance variability in a large semi-arid catchment down ward approach to hydrological model development. Journal of Hydrology, 254(1-4), 174-198.

Karssenberg, D., Burrough, P.A., Sluiter, R., and De Jong, K. (2001). Review PCRaster software and course materials for teaching numerical modelling in the environmental sciences, Transactions in GIS 5 (2), 99–110.

Kebede, S., Travi, Y., Alemayehu, T., Ayenew, T., and Aggarwal, P. (2003). Tracing sources of recharge to ground waters in the Ethiopian Rift and bordering plateau: Isotopic evidence,

Paper presented at the Fourth International Conference on Isotope for Groundwater Management, Vienna, IAEA, 2003.

Kebede, S. (2004). Approaches isotopique et geochimique pour l'etudedes eaux souterraines et des lacs: Exemples du haut bassin du Nil Bleu et du rift Ethiopien [Environmental isotopes and geochemistry in groundwater and lake hydrology: cases from the Blue Nile basin, main Ethiopian rift and Afar, Ethiopia], unpublished PhD thesis, University of Avignon, France, 2004, 162p.

Kebede, S., Travi, Y., Alemayehu, T., and Marc, V. (2006). Water balance of Lake Tana and its sensitivity to fluctuations in rainfall, Blue Nile basin, Ethiopia Journal of Hydrology 316 233–247.

Kebede, S., and Travi, Y. (2012). Origin of the $\delta^{18}O$ and $\delta^2H$ composition of meteoric waters in Ethiopia. Quaternary International, 257, 4-12, doi:10.1016/j.quaint.2011.09.032.

Kendall, M.G. (1975). Rank correlation Methods, Charles Griffin, London.

Kendall, C., and Caldwell, E.A. (1998). Fundamentals of isotope geo-chemistry, in: Isotope Tracers in Catchment Hydrology, edited by: Kendall, C and McDonnell, J.J., Elsevier Science, Amsterdam, 51-86, 1998.

Kendall, C., and Coplen, T.B. (2001). Distribution of oxygen -18 and deuterium in river waters across the United States. Hydrol.Process. 15, 1363-1393, doi 10.1002/hyp.217, 2001.

Kim, U., Kaluarachchi, J.J., and Smakhtin, V.U.M. (2008). Generation of Monthly precipitation under climate change for the upper Blue Nile River Basin, Ethiopia. J.Am. Water Resources, JAWRA 44(5):1231-1247. DOI: 10.1111/j.1752-1688.2008.00220.x.

Kim, U., and Kaluarachchi, J.J. (2008). Application of parameter estimation and regionalization methodologies to ungauged basins of the upper Blue Nile river basin, Ethiopia. Journal of hydrology 362, 39-56.

Kirchner, J.W. (2006). Getting the right answers for the right reasons: linking measurements, analyses, and models to advance the science of hydrology, Water Resour. Res., 42: W03S04, doi: 10.1029/2005WR004362.

Kirchner, J.W., Tetzlaff, D., and Soulsby, C. (2010). Comparing chloride and water isotopes as hydrological tracers in two Scottish catchments. Hydrol. Process. 24, 1631-1645, doi: 10.1002/hyp.7676.

Klemes, V. (1983). Conceptualization and Scale in Hydrology. Journal of Hydrology, 65(1-3),1-23.

Kloos, H., and Legesse, W. (2010). Water Resources Management in Ethiopia: Implications for the Nile Basin. ISBN 978-1-60497-665-6, Cambria press.

Lambin, E.F., Geist, H.J., and Lepers, E. (2003). Dynamics of land use and land cover change in tropical regions. Annu.Rev. Environ. Resour., 28: 205-41.

Laudon, H., Sjöblom, V., Buffam, I., Seibert, J., and Mörth, M. (2007). The role of catchment scale and landscape characteristics for runoff generation of boreal stream, J. Hydrol., 344, 198-209 doi:10.1016/j.jhydrol.

Legates, D.R., and McCabe, G.J. (1999). Evaluating the use of "goodness- of- fit" measures in hydrologic and hydro-climatic model validation. Water Resources research 35, 233-241.

Lehmann, P., Hinz, C., McGrath, G., Tromp-van Meerveld, H.J., and McDonnell, J.J. (2007). Rainfall threshold for hillslope outflow: an emergent property of flow pathway connectivity. Hydrology and Earth System Sciences, 11, 1047–1063.

Levin, N.E., Zipser, E.J., and Cerling, T.E. (2009). Isotopic composition of waters from Ethiopia and Kenya: Insight into moisture sources for eastern Africa. Journal of Geophysical Research., Vol. 114, D23306, doi: 10.1029/2009JD0121669.

LindstrÖm, G., Johansson, B., Persson, M., Gardelin, M., and Bergstr¨om, S. (1997). Development and test of the distributed HBV-96 hydrological model. Journal of Hydrology, 201(1-4), 272-288.

Liu, Y., Fan, N., An, S., Bai, X., Liu, F., Xu, Z., Wang, Z., and Liu, S. (2008a). Characteristics of water isotopes and hydrograph separation during the wet season in the Heishui River, China, J. hydrol., 353, 314-321.

Liu, B.M, Collick, A.S, Zeleke, G, Adgo E, Easton, Z.M., and Steenhuis, T.S. (2008b). Rainfall-discharge relationships for a monsoonal climate in the Ethiopian highlands. Hydrological Processes 22: 1059–1067.

Lorup, J.k., Refsgaard, J.C., and Mazvimavi, D. (1998). Assessing the effect of land use change on catchment runoff by combined use of statistical tests and hydrological Modelling: case studies from Zimbabwe. Journal of Hydrology 205:147-163.

Love, D., Uhlenbrook, S., Twomlow, S., and Van der Zaag, P. (2010). Changing hydro-climatic and discharge patterns in the northern Limpopo Basin, Zimbabwe. ISSN 0378-4738 (Print) Water SA Vol. 36 No. 3.

Maloszewski, P., and Zuber, A. (1982). Determining the turnover time of groundwater systems with the aid of environmental tracers, Models and their applicability, J. Hydrol., 57, 207–231.

Mann, H.B. (1945). Nonparametric tests against trend, Econometrica, 13, 245-259.

Masih, I., Uhlenbrook, S., Maskey, S., and Smakhtin, V. (2010). Stream flow trends and climate linkages in the Zagros Mountains, Iran. Climatic Change, 104: 317-338, doi 10.1007/s10584-009-9793-x.

McDonnell, J.J., Bonell, M., Stewart., M.K., and Pearce, A.J. (1990). Deuterium variations in storm rainfall-Implications for stream hydrograph separation, Water Resour. Res., 26: 455–458.

McDonnell, J.J., Stewart, M.K., and Owens, I.F. (1991). Effect of catchment scale subsurface mixing on stream isotopic response, Water Resour. Res., 27, 3065-3073.

McGlynn, B, McDonnell, J, Stewart, M, and Seibert, J. (2003). On the relationship between catchment scale and stream water mean residence time. Hydrological Processes, 17: 175–181.

McGlynn, B.L., McDonnell, J.J., Seibert, J, and Kendall, C. (2004).Scale effects on headwater catchment runoff timing, flow sources and groundwater–streamflow relations. Water Resources Research, 40, W07504.

McGuire, K.J., DeWalle, D.R., and Gburek, W.J. (2002). Evaluation of mean residence time in subsurface waters using oxygen-18 fluctuations during drought conditions in the semi-Appalachains, J. Hydrol., 261 (2002) 132-149.

McGuire, K.J., McDonnell, J.J., Weiler, M., Kendall, C., McGlynn, C.L., Welker, J.L., and Siebert, J. (2005). The role of topography on catchment scale water residence time, Water Resour. Res., 41, W05002, doi: 10.1029/2004WR003657.

McGuire, K.J., and McDonnell, J.J. (2006). A review and evaluation of catchment transit time modelling, J. Hydrol., 330, 543–563.

Milly, P.C.D. (1994). Climate, Soil water storage, and the average annual water balance. Water Resources Research Vol.30, No.7, Pages 2143-2156.

Mishra, A., and Hata, T. A. (2006). Grid based runoff generation and flow routing model for the upper Blue Nile Basin, Hydrological Sciences J., 51 (2), pp 191-205.

Mohamed, Y.A., Van den Hurk, B. J. J. M., Savenije, H.H.G., and Bastiaanssen, W.G.M. (2005). Hydro-climatology of the Nile: results from a regional climate model, Hydrol. Earth Syst. Sci., 9, 263–278, doi: 10.5194/hess-9-263.

Montanari, L., Sivapalan, M., and Montanari, A. (2006). Investigation of dominant hydrological processes in a tropical catchment in a monsoonal climate via downward approach. Hydrological Earth System, Sci., 10,769-782.

Moore, G.W., and Heilman, J.L. (2011). Eco-hydrology Bearings—Invited Commentary. Proposed principles governing how vegetation changes affect transpiration. Eco-hydrology 4: 351–358. doi: 10.1002/eco.232.

Moraes, J.M., Pellegrino, G.Q., Ballester, M.V., Martinelli, L.A., Victoria, R.L., and Krusche, A.V. (1998). Trends in hydrological parameters of a southern Brazilian watershed and its relation to human induced changes. Water Resources Management 12, 295–311.

Mosley, M.P. (1979). Streamflow Generation in a Forested Watershed, New-Zealand. Water Resources Research, 15(4), 795-806.

Mu, X., Zhang, L., McVicar, T.R., Chille, B., and Gau, P. (2007). Analysis of the impact of conservation measures on stream flow regime in catchments of the Loess Plateau, China. Hydrological Processes, 21, 2124–2134.

Mul, M. L., Mutiibwa, K. R., Uhlenbrook, S, and Savenije, H.H.G. (2008). Hydrograph separation using hydro-chemical tracers in the Makanya catchment, Tanzania, Phys. Chem. Earth., 33, 151–156.

Munyaneza, O., Wenninger, J., and Uhlenbrook, S. (2012). Identification of runoff generation processes using hydrometric and tracer methods in a meso-scale catchment in Rwanda, Hydrol. Earth Syst. Sci., 16, 1991–2004, doi: 10.5194/hess-16-1991.

Nash, J. E., and Sutcliffe, J. V. (1970). River flow forecasting through conceptual models, Part I –A discussion of principles, J. Hydrol., 10, 282–290.

Nawaz, R., Bellerby, T., Sayed, M., and Elshamy M. (2010). Blue Nile runoff sensitivity to climate change. The Open Hydrology Journal, 4, 137-151.

NMSA (National Meteorological service Agency). (1996), Climatic and agro-climatic resources of Ethiopia, NMSA Meteorological Research Report Series. V1, No. 1, Addis Ababa, 137p.

Ol'dekop, E.M. (1911). On evaporation from the surface of river basins, Transactions on Meteorological Observations. Lur-evskogo, Univ.of Tartu, Tartu, Estonia.

Orr, H.G., and Carling, P.A. (2006). Hydro-climatic and land use changes in the river Lune catchment, North West England, implications for catchment management. River Res. Appl. 22,239–255, doi: 10.1002/rra.908.

Ott, B., and Uhlenbrook, S. (2004). Quantifying the impacts of land use changes at the event and seasonal time scales using a process oriented catchment model. Hydrology and Earth system Science 8(1), 62-78.

Pagano, T.P., Garen, D., Sorooshian, S. (2004). Evaluation of official Western US seasonal water supply outlooks, 1922–2002. Journal of Hydrometeorology, 5, 896–909.

Pearce, A.J., Stewart, M.K., and Sklash, M.G. (1986). Storm runoff generation in humid headwater catchments. Where does the water come from? Water Resour. Res., 22: 1263-1272.

Pettitt, A.N. (1979). A non-parametric approach to the change-point problem. Appl. Statistics 28, 126–135.

Phillips, R.W., Spence, C., and Pomeroy, J.W. (2011). Connectivity and runoff dynamics in heterogeneous basins. Hydrological Processes, 25, 3061–3075.

Pike, J.G. (1964). The estimation of annual runoff from meteorological data in a tropical climate. Journal of Hydrology, 2, 116-123.

Potter, Nicholas. J., and Zhang, Lu. (2009). Inter annual variability of catchment water balance in Australia. Journal of Hydrology , 369, 120-129.

Richter, B., Baumgartner, J.V., Powell, J., and Braun, D.P. (1996). A method for assessing Hydrologic alteration within the ecosystems. Conservation Biology, pages 1163-1174, Volume 10, No.4.

Rientjes, T.H.M., Perera, B.U.J., Haile, A.T., Reggiani, P., and Muthuwatta, L.P. (2011a). Regionalization for lake level simulation - the case of Lake Tana in the Upper Blue Nile, Ethiopia, Hydrol. Earth Syst. Sci., 15, 1167–1183, doi: 10.5194/hess-15-1167.

Rientjes, T.H.M., Haile, A.T., Kebede, E., Mannaerts, C.M.M., Habib, E., and Steenhuis, T.S. (2011b). Changes in land cover, rainfall and stream flow in Gilgel Abbay catchment, Upper Blue Nile basin–Ethiopia. Hydrol. Earth Syst. Sci., 15, 1979–1989, 2011, doi: 10.5194/hess-15-1979.

Rodgers, P, Soulsby, C., Waldron, S., and Tetzlaff, D. (2005a). Using stable isotope tracers to access hydrological flow paths, residence times and landscape influences in a nested meso-scale catchment, Hydrol. Earth Syst. Sci., 9: 139–155.

Rodgers, P, Soulsby, C., and Waldron, S. (2005b). Stable isotope tracers as diagnostic tools in up scaling flow path understanding and residence time estimates in a mountainous meso-scale catchment. Hydrol. Process. 19, 2291–2307.

Rozanski, K., Araguas-Araguas, L., and Gonfiantini, R. (1996). Isotope patterns of precipitation in the east African region, In: Johnson, T.C., Odada, E. (Eds.), The Liminology, Climatology and Paleo-climatology of the East African Lakes, Gordon and Breach, Toronto, pp. 79-93.

Sankarasubramania, A., and Vogel, R.M. (2003). Hydro-climatology of the continental United States. Geophysical Research Letters Vol.30, No., 1363, doi: 0.1029/2002GL015937.

Savenije, H.H.G. (2001). Equifinality a blessing in disguise? Hydrological processes, 15, 2835-2838, doi:10.1002/hyp.494.

Savenije, H.H.G. (2004). The importance of interception and why we should delete the term evapotranspiration from our vocabulary. Hydrological Processes, 18(8), 1507-1511.

Savenije, H.H.G. (2009). The art of hydrology. Hydrol. Earth Syst. Sci., 13, 157–161

Savenije, H.H.G. (2010). Topography driven conceptual modelling (FLEX-Topo). Hydrol. Earth Syst. Sci., 14, 2681–2692, doi: 10.5194/hess-14-2681.

Schaefli, B. and Gupta, H.V. (2007). Do Nash values have value? Hydrological Processes, 21(15): 2075-2080.

Schreiber, P. (1904). Über die Beziehungen zwischen dem Niederschlag und der Wasserführung der Flüsse in Mitteleuropa. Meteorol. Z., 21, 441–452.

Seibert, J., and McDonnell, J.J. (2002). On the dialog between experimentalist and modeller in catchment hydrology: Use of soft data for multi-criteria model calibration, Water Resour. Res., 38(11), 1241, doi: 10.1029/2001WR000978.

Seibert, J., Bishop, K., Rodhe, A., and McDonnell, J.J. (2003). Groundwater dynamics along a hillslope: a test of the steady state hypothesis. Water Resources Research, doi: 1029/2002WR001404, 39 (1), 1014.

Seibert, J., and McDonnell, J.J. (2010). Land-cover impacts on streamflow: a change-detection modelling approach that incorporates parameter uncertainty. Hydrological Sciences Journal 55: 3, 316-332.

Seleshi, Y., and Zanke, U. (2004). Recent change in rainfall and rainy days in Ethiopia. International Journal of climatology Int. J. Climatol., 24: 973–983, doi: 10.1002/joc.1052.

Seleshi, Y., Camberlin, P. (2006). Recent in dry spell and extreme rainfall events in Ethiopia. Theor. Appl. Climatol. 83, 181–191, doi.org/10.1007/s00704-005-0134-3.

Setegne, S.G., Srinivasan, R., Melese, A.M., and Dargahi, B. (2010). SWAT model application and 612 prediction uncertainty analysis in the Lake Tana Basin, Ethiopia. Hydrological processes Hydrol. Process., 24, 357–367.

Shao, Q., Zhang, L., Chen, Y., Singh, V.P. (2009). A new method for modelling flow duration curves and predicting streamflow regimes under altered land-use conditions / Une nouvelle méthode de modélisation des courbes de debits classés et de prévision des régimes d'écoulement sous conditions modifiées d'occupation du sol'. Hydrological Sciences Journal, 54: 3, 606-622.

Shanley, J.B., Kendall, C.T., Smith, E D., Wolock., and J.J. McDonnell. (2002). Controls on old and new water contributions to stream flow at some nested catchments in Vermont, USA, Hydrol. Processes., 16, 589–609.

Sieber, A., and Uhlenbrook, S. (2005) Sensitivity analyses of a distributed catchment model to verify the model structure. Journal of Hydrology, volume 310, issues 1-4, page 216-235.

Singh, V.P. (1995). Watershed modelling. In: V.P. Singh (ed.), Computer Models of Watershed Hydrology. Water Resources Publication, Highlands Ranch, CO, USA, pp. 1-22.

Siriwardena, L., Finlayson, B.L., and McMahon, T.A. (2006). The impact of land use change on catchment hydrology in large catchments: The Comet River, Central Queensland, Australia. Journal of Hydrology, 326: 199–214.

Sivakumar, B. (2008). Dominant processes concept, model simplification and classification framework in catchment hydrology, Stoch Environ Res Risk Assess., 22:737–748, doi 10.1007/s00477-007-0183-5.

Sivapalan, M. (2003a). Process complexity at hillslope scale, process simplicity at the watershed scale: Is there a connection? Hydrological Processes 17: 1037–1041.

Sivapalan, M. (2003b). Prediction of ungauged basins: a grand challenge for theoretical hydrology. Hydrological Processes, 17(15), 3163-3170.

Sivapalan, M., Blo¨schl, G., Zhang, L., and Vertessy, R. (2003c). Downward approach to hydrological prediction, Hydrol. Processes, 17, 2101–2111.

Sivapalan, M., Takeuchi, K. and Franks, S.W. (2003d). IAHS Decade on Predictions in Ungauged Basins (PUB), 2003-2012: Shaping an exciting future for the hydrological sciences. *Hydrol. Sci. J.,* 48(6): 857-880.

Sivapalan, M., and Young, P. C. (2005). Downward approach to hydrological model development. Encyclopaedia of Hydrological Sciences.

Sivapalan, M. (2005). Pattern, Process and Function: Elements of a Unified Theory of Hydrology at the Catchment Scale. In: Encyclopedia of Hydrological Sciences, M. G. Anderson (Managing Editor), J. Wiley & Sons, in Press.C.

Sivapalan, M. (2009). The secret to doing better hydrological science: change the question! Hydrological Processes, 23: 1391–1396.

Sklash, M.G., and Farvolden, R.N. (1979). The role of groundwater in storm runoff. Journal of Hydrology, 43 (1979) 45-65.

Solomatine, D.P. (2011). Hydrological Modelling. Treatise on Water Science, 2: 435–457.

Son, K., and Sivapalan, M. (2007). Improving model structure and reducing parameter uncertainty in conceptual water balance models through the use of auxiliary data. Water Resources Research, Vol 43, W01415, doi: 10.1029/2006WR005032.

Soulsby, C, Malcolm, R, Helliwell, R, Ferrier, R.C., and Jenkins, A. (2000). Isotope hydrology of the Allt a' Mharcaidh catchment, Cairngorms, Scotland: implications for hydrological pathways and residence times, Hydrol. Processes., 14: 747–762.

Soulsby, C., Tetzlaff, D., Dunn, S.M., Waldron., and S. (2006). Scaling up and out in runoff process understanding: insight from nested experimental catchment studies, Hydrol. Processes., 20:2461-2465.

Soulsby, S., and Tetzlaff, D. (2008). Towards simple approaches for mean residence time estimation in ungauged basins using tracers and soil distributions, J., Hydrol. 363, 60–74.

Spence, C. and Woo, M.K. (2006). Hydrology of sub-arctic Canadian Shield: heterogeneous headwater basins. Journal of Hydrology, 317, 138–154.

Spence, C., Guan, X. J., Phillips, R., Hedstrom, N., Granger, R., and Reid, B. (2010). Storage dynamics and streamflow in a catchment with a variable contributing area. Hydrological Processes, 24, 2209–2221.

Steenhuis, T., Collick, A., Easton, Z., Leggesse, E., Bayabil, H., White, E., Awulachew, S., Adgo, E., and Ahmed, A. (2009). Predicting discharge and sediment for the Abay (Blue Nile) with a simple model, Hydrol. Process., 23, 3728-3737.

Stevens, J. C. (1907). "A method of estimating stream discharge from a limited number of gaugings." Engineering News, Vol. 58, No. 3. pp. 52-53.

Su, Z. (2002). The Surface Energy Balance System (SEBS) for estimation of turbulent heat fluxes. Hydrology and Earth System Sciences, 6(1): 85–99.

Sullivan, A., Ternan, J.L., and Williams, A.G. (2004). Land use change and hydrological response in the Camel catchment, Cornwall. Applied Geography, 24: 119–137.

Sugawara, M. (1967). The flood forecasting by a series storage type model, Proc. of International Symposium on floods and their computation, Leningrad, USSR, IAHS Publication, 85: 1-6.

Sutcliffe, J. V., and Y. P. Parks. (1999). The Hydrology of the Nile, IAHS Special Publication no. 5, IAHS Press, Institute of Hydrology, Wallingford, Oxfordshire OX10 8BB, UK.

Taylor, C.B., Wilson, D.D., Borwn, L.J., Stewart, M.K., Burdon, R.J., and Brailsford, G.W. (1989). Sources and flow of North Canterbury plains ground water, New Zealand, J. Hydrol., 106: 311-340, 1989.

Teferi, E., Uhlenbrook, S., Beweket, W., Wenninger, J., and Simane, B. (2010). The use of remote sensing to quantify wet land loss in the Choke Mountain range, upper Blue Nile Ethiopia. Hydrol. Earth Syst. Sci., 14, 2415–2428, doi: 10.5194/hess-14-2415.

Teferi, E., Bewket, W., Uhlenbrook, S., and Wenninger, J. (2013). Understanding recent land use and land cover dynamics in the source region of the Upper Blue Nile, Ethiopia: Spatially explicit statistical modelling of systematic transitions. Agriculture, Ecosystem and Environment 165 (2013) 98-117.

Tekleab, S., Uhlenbrook, S., Mohammed, Y., Savenije, H.H.G., Temesgen, M., and Wenninger, J. (2011). Water balance modelling of the upper Blue Nile catchments using a top-down approach. Hydrology and Earth System Science, 15: 2179–2193, doi: 10.5194/hess-15-2179.

Tekleab, S., Mohamed, Y., and Uhlenbrook, S. (2013) Hydro-climatic trends in the Abay/Upper Blue Nile basin, Ethiopia. Journal of Physics and Chemistry of the Earth, doi: 10.1016/j.pce. 2014.04.017.

Tekleab, S., Mohamed, Y., Uhlenbrook, S., and Wenninger, J. (2014a). Hydrologic responses to land cover change, the case of Jedeb mesoscale catchment, Abay/Upper Blue Nile basin, Ethiopia, Hydrol. Process., doi:10.1002/hyp.9998, 28, 5149-5161.

Tekleab, S., Wenninger, J., and Uhlenbrook, S. (2014b). Characterisation of stable isotopes to identify residence times and runoff components in two meso-scale catchments, Abay/Upper Blue Nile basin, Ethiopia. Hydrol. Earth Syst. Sci., 18, 2415–2431, doi: 10.5194/hess-18-2415.

Temesgen, M., Uhlenbrook, S., Belay, S., Vander Zaag, P., Mohamed, Y., Wenninger, J., and Savenije, H.H.G. (2012). Impacts of conservation tillage on the hydrological and agronomic performance of fanya juus in the upper Blue Nile (Abbay) river basin. Hydrology and. Earth System Science Hydrol. Earth Syst. Sci., 16, 4725–4735, doi: 10.5194/hess-16-4725.

Tesemma, Z.K, Mohamed, Y, A., and Steenhuis T.S. (2010). Trends in rainfall and runoff in the Blue Nile Basin: 1964–2003. Hydrological Processes, doi: 10.1002/hyp.7893.

Tetzlaff, D., Waldron, S., Brewer, M.J., and Soulsby, C. (2007a). Assessing nested hydrological and hydo-chemical behaviour of a meso-scale catchment using continuous tracer data, J. Hydrol., 336, 430-443.

Tetzlaff D., Soulsby, C., Waldron, S., Malcolm, I.A., Bacon, P.J, Dunn, S.M., and Lilly, A. (2007b). Conceptualization of runoff processes using GIS and tracers in a nested meso-scale catchment, Hydrol. Processes, 21: 1289–1307.

Tetzlaff, D., Seibert, J., and Soulsby, C. (2009). Inter-catchment comparison to assess the influence of topography and soils on catchment transit times in a geomorphic province; in a Cairngorm Mountains Scotland, Hydrol. Processes., 23:1874-1886.

Tilahun S.A., Guzman, C.D., Zegeye, A.D., Engda, T.A., Collick, A.S., Rimmer, A. and Steenhuis, T.S. (2013a) An efficient semi-distributed hillslope erosion model for the sub humid Ethiopian Highlands. *Hydrology and Earth System Sciences* 17:1051–1063.

Tilahun S.A., Mukundan, R., Demisse, B.A., Engda, T.A., Guzman, C.D., Tarakegn, B.C., Easton, Z.M., Collick, A.S., Zegeye, A.D., Schneiderman, E.M., Parlange, J.-Y. and Steenhuis, T.S. (2013b) A saturation excess erosion model. *Transactions of the ASABE* 56:681–695.

Tilahun, S.A., Guzman, C.D., Zegeye, A, D., Dagnew, D.C., Collic, A.S., Yitaferu, B., and Steenhuis, T.S. (2014) Distributed discharge and sediment concentration predictions in the sub-humid

Ethiopian highlands: the Debre Mawi catchment. Hydrological processes, doi: 10.1002/hyp.10298, in press.

Thornthwaite, C.W, Mather, J.R. (1955) The Water Balance. Publication 8.

Tolba, M.K., and El-Kholy, O.A. (Eds.). (1992). The World Environment 1972–1992: Two Decades of Challenge, Chapman & Hall, London.

Tromp-Van-Meerveld, H. J., and McDonnell, J. J. (2006). Threshold relations in subsurface stormflow 1. A storm analysis of the Panola hillslope, Water Resour. Res. 42, W02410, doi: 10.1029/2004WR003778.

Turc, L. (1954). Le bilan d'eau des sols Relation entre la precipitation l'evapporation et l'ecoulement,. Ann. Agron., 5, 491–569.

Uhlenbrook, S., Seibert J., Leibundgut Ch., and Rodhe, A. (1999). Prediction uncertainty of conceptual rainfall-runoff models caused by problems to identify model parameters and structure. Hydrological Sciences Journal, 44, 5, 279-299.

Uhlenbrook, S., and Leibundgut, C.h, (2002). Process-oriented catchment modelling and multiple-response validation, Hydrol. Process., 16, 423–440.

Uhlenbrook, S., Frey, M., Leibundgut, C., and Maloszewski, P. (2002). Hydrograph separations in a meso-scale mountainous basin at event and seasonal time scales, Water Resour. Res., 38 (6), 31, 1-13.

Uhlenbrook, S., and Hoeg, S. (2003). Quantifying uncertainties in tracer based hydrograph separations: a case study for two, three and five component hydrograph separations in a mountainous catchment, Hydrol. Processes., 17: 431-453.

Uhlenbrook, S., Roser, S., and Tilch, N. (2004). Hydrological process representation at the meso-scale: the potential of a distributed, conceptual catchment model. Journal of Hydrology 291: 278–296.

Uhlenbrook, S., Wenninger, J. (2006). Identification of flow pathways along hillslopes using electrical resistivity tomography (ERT). In M. Sivapalan et al. (Eds.), Predictions in ungauged basins: promise and progress (Vol. 303, p. 15-20). England.

Uhlenbrook, S., Didszun, J. and Wenninger, J. (2008). Sources areas and mixing of runoff components at the hillslope scale – a multi-technical approach. Hydrological Sciences Journal, 53(4).

Uhlenbrook, S., Mohamed, Y., and Gragne, S. (2010). Analyzing catchment behaviour through catchment 630 modelling in the Gilgel Abay, Upper Blue Nile River Basin, Ethiopia. Hydrol. Earth Syst. 631 Sci., 14, 2153–2165, doi: 10.5194/hess-14-2153.

UNESCO (2004). United Nations, Educational, Scientific and Cultural Organization. National Water Development Report for Ethiopia, UN-WATER / WWAP/2006/7, World Water Assessment program, Report, MOWR, Addis Ababa, Ethiopia.

Van der Velde Y., de Rooij, G.H., Rozemeijer, J.C., Van Geer, F.C., and Broers, H.P. (2010). Nitrate response of a lowland catchment: On the relation between stream concentration and travel time distribution dynamics. Water Resour. Res., 46, W11534, doi: 10.1029/2010WR009105.

Vertesy, R., Hatton, T., O'haughnesy, P., and Jayasuriya, M. (1993). Predicting water yield from a mountain ash forest using a terrain analysis based catchment model. Journal of Hydrology 150; 665-700.

Viste, E., and Sorteberg, A. (2013). Moisture transport into the Ethiopia highlands, Int. J. Climatol., 33:249-263.

Vrugt, J. A., Gupta, H.V., Bastidas, L.A., Bouten, W., and Sorooshian, S. (2003). Effective and efficient algorithm for multi-objective optimization of hydrologic models. Water Resources Research, 39(8), 1214.

Wagener, T., Boyle, D.P., Lees, M.J., Wheater, H.S., Gupta, H.V., and Sorooshaian, S. A. (2001). Framework for development and application of hydrological models. Hydrology and Earth System Sciences, 5(1), 13–26.

Wagener, T., McIntyre, N., Lees, M. J., Wheater, H. S., and Gupta, H. V. (2003). Towards reduced uncertainty in conceptual rainfall-runoff modelling: dynamic identifiably analysis, Hydrol. Process., 17, 455–476, doi:10.1002/hyp.1135.

Wang, G.X., Liu, J.Q., Kubota, J., and Chen, L. (2007b). Effect of land-use changes on hydrological processes in the middle basin of the Heihe River, northwest China. Hydrological Processes 21: 1370–1382, doi: 10.1002/hyp.6308.

Wang, Y., Dietrich, J., Voss, F., and Pahlow, M. (2007a). Identifying and reducing model structure uncertainty based on analysis of parameter interaction. Adv. Geosci., 11, 117–122.

Wels, C., Cornett, R. J., and Lazerte, B.D. (1991). Hydrograph separation: a comparison of geochemical and isotopic tracers, J. Hydrol., 122, 253–274.

Wenninger, J., Uhlenbrook, S., Lorentz, S. and Leibundgut, C. (2008). Identification of runoff generation processes using combined hydrometric, tracer and geophysical methods in a headwater catchment in South Africa. Journal of Hydrological Sciences, 53(1), 65-80.

Wesseling, C.G., Karssenberg, D.J., Burrough, P.A., and Van Deursen, W.P.A. (1996). Integrated dynamic environmental models in GIS: The development of a Dynamic Modelling language, Transactions in GIS, 11, 4048.

White, E. D., Easton, Z. M., Fuka, D. R., Collick, A. S., Adgo, E., McCartney, M., Awulachew, S. B., Selassie, Y., and Steenhuis, T. S. (2010) Development and application of a physically based landscape water balance in the SWAT model, Hydrol. Process., 23, 3728–3737, doi:10.1002/hyp.7876.

Winsemius, H.C., Schaefli, B., Montanari, A., and Savenije, H.H.G. (2009). On the calibration of hydrological models in ungauged basins: A framework for integrating hard and soft hydrological information. Water Resources Research, VOL. 45, W12422, doi: 10.1029/2009WR00706.

Winsemius, H.C. (2009). Satellite data as complementary information for hydrological modelling. Technical University of Delft, PhD. thesis Pp186. VSSD Leeghwaterstraat 42, 2628 CA Delft, The Netherlands.

Wissmeier, L., and Uhlenbrook, S. (2007). Distributed, high-resolution modelling of 18O Signals in a meso-scale catchment, Journal of Hydrology 332, 497–510, doi:10.1016/j.jhydrol.2006.08.003.

Yang, D., Sun, F., Liu, Z., Cong, Z., Ni, G., and Lei, Z. (2007). Analyzing spatial and temporal variability of annual water-energy balance in non-humid regions of China using the Budyko hypothesis. Water Resources Research, VOL. 43, W04426, doi: 10.1029/2006 WR005224.

Yang, D., Shao, W., Yeh, P.J.F., Yang, H., Kanae, S., and Oki, T. (2009). Impact of vegetation coverage on regional water balance in the non-humid regions of china. Water Resources Research VOL. 45, W00A14, doi: 10.1029/2008WR006948.

Yang, Z., Zhou, Y., Wenninger, J., and Uhlenbrook, S. (2012). The cause of flow regime shift in the semi- arid Humiliate River Northwest China. Hydrology and. Earth System Science, 16: 87–103, doi: 10.5194/hess-16-87.

Yue S, Pilon P, Phinney B, and Cavadias G. (2002). The influence of autocorrelation on the ability to detect trend in hydrological series. Hydrological Processes, 16: 1807–1829.

Yue, S., Pilon, P., and Phinney, B. (2003). Canadian stream flow trend detection: impacts of serial and cross-correlation, Hydrolog. Sci.J., 48(1), 51–63.

Yue, S., and Pilon, P. (2004). A comparison of the power of the t test, Mann Kendall and bootstrap tests for trend detection. Hydrological Sciences Journal, 49(1), 21–38.

Zhe, E., and Sivapalan, M. (2009). Threshold behaviour in hydrological systems as (human) geo-ecosystems: manifestations, controls, implications. Hydrol. Earth Syst. Sci., 13, 1273–1297.

Zeleke, G., and Hurni, H. (2001). Implications of land use and land cover dynamics for mountain resource degradation in the North-western Ethiopian highlands. Mountain Research and Development. Vol 21 No 2:184-191.

Zhang, L., Dawes, W.R., and Walker, G.R. (1999). Predicting the effect of vegetation changes on catchment average water balance. Tech. Rep.99/12 Coop. Res. Cent. Catch. hydrol. Canbera, ACT.

Zhang, L., Dawes, W.R., and Walker, G.R. (2001a). Response of mean annual evapotranspiration to vegetation changes at catchment scale. Water Resources Research 37, No.3, Pages 701-708.

Zhang, L., Hickel, K., Dawes, W.R., Chiew, F.H.S., and Western, A.W. (2004). A rational function approach for estimating mean annual evapotranspiration. Water Resources Research, VOL. 40, W02502, doi: 10.1029/2003WR002710.

Zhang, L., Potter, N., Hickel, K., Zhang, Y.Q., and Shao, Q.X. (2008). Water balance modelling over variable time scales based on the Budyko framework-Model development and testing. Journal of Hydrology, 360, 117–131.

# References

Zhang, G.P., and Savenije, H.H.G. (2005). Rainfall-runoff modelling in a catchment with a complex groundwater flow system: application of the Representative Elementary Watershed (REW) approach. Hydrology and Earth System Sciences, 9(3), 243-261.

Zhang, X., Harvey, K.D., Hogg, W.D., and Yuzyk, T.R. (2001b). Trends in Canadian streamflow. Water Resour. Res., 37:987–998.

Zhao, C., Nan, Z., and Cheng, G. (2005). Evaluating Methods of Estimating and Modelling Spatial Distribution of Evapotranspiration in the Middle Heihe River Basin, China. American Journal of Environmental Sciences, 1 (4), 278-285.

## Appendix A

Table A1. Description of stream flow gauging stations in the Abay/Upper Blue Nile basin. Numbers in brackets in column one indicate the location of the catchments referring to figure 3.1. Percentage missing refers to the number of missing data records in the given period.

| River / Catchment name | Catchment area (km²) | Mean annual stream flow (m³/s) | CV (%) | Stream gauging location | Period of record | Daily regression coefficient (r²) | Remark |
|---|---|---|---|---|---|---|---|
| Rib (1) | 1289 | 12.52 | 30 | Near Addis Zemen | 1973-2003 | Rib with Gumera (0.71) | 0.6 % missing data and 1982 data ignored in the analysis |
| Gumera (2) | 1269 | 28.25 | 28 | Near Bahir Dar | 1973-2006 | 0.71 | 0.94 % |
| Koga (3) | 295 | 4.4 | 28 | In Merawi | 1973-2005 | Koga with Gilgel Abay (0.67) | 0.77 % missing data, and 1981 and 1982 data ignored in the analysis |
| Gilgel Abay (4) | 1659 | 51.21 | 11 | Near Merawi | 1973-2005 | 0.67 | 0.33 % missing data |
| Jedeb (5) | 277 | 7.7 | 30 | In Yewla | 1973-2010 | - | 4.87 % missing data. All ignored in the analysis |
| Chemoga (6) | 358 | 5.45 | 31 | Near Debre Markos | 1973-2010 | - | 4.39 % missing data. All ignored in the analysis |
| Guder (7) | 512 | 11.23 | 17 | In Guder | 1973-2006 | - | - |
| Muger (8) | 486 | 7.02 | 24 | Near Chancho | 1973-2004 | - | 0.64 % missing data. Ignored in the analysis |
| Neshi (9) | 327 | 6.09 | 38 | Near Shambu | 1973-2004 | - | - |

Table A2. Description of meteorological gauging stations in the Abay/Upper Blue Nile basin.

| Meteorological station | mean annual rainfall (mm a$^{-1}$) | CV (%) for Rainfall data | CV(%) for Temperature data | mean annual temperature (°c) | Elevation (m.a.s.l) | Period of record for rainfall data | Period of record for temperature data | % missing | |
|---|---|---|---|---|---|---|---|---|---|
| | | | | | | | | Rainfall record | Average temperature record |
| Addis Ababa (Bole) | 1069 | 16 | - | - | 2354 | 1954-2004 | - | - | - |
| Gonder | 1085 | 18 | - | - | 1967 | 1973-2001 | - | 0.89 | - |
| Debre Tabor | 1536 | 20 | - | - | 2690 | 1973-2003 | - | 6.11 | - |
| Nedjo | 1557 | 20 | - | - | 1800 | 1973-2003 | - | 12 | - |
| Bahir Dar | 1405 | 17 | 5 | 19.3 | 1770 | 1961-2007 | 1963-2006 | 0.5 | 4.2 |
| Debre Markos | 1298 | 12 | 3 | 16.1 | 2515 | 1954-2010 | 1963-2006 | 0.59 | 1.6 |
| Dessie | 1196 | 16 | - | - | 2500 | 1973-2004 | - | 3.2 | - |
| Alem Ketema | 1120 | 16 | 5 | 18.9 | 2280 | 1973-2003 | 1974-2002 | 10 | 6.5 |
| Fincha | 1322 | 12 | 2 | 22.7 | 2320 | 1980-2008 | 1980-2008 | - | - |
| Jimma | 1481 | 16 | 16 | - | 1700 | 1973-2004 | - | 5.8 | - |
| Bedelle | 1878 | 21 | 7 | 18.6 | 2030 | 1973-2003 | 1973-2004 | 8.1 | 9.2 |
| Nekemet | 2067 | 14 | 5 | 16.5 | 2080 | 1971-2003 | 1985-2005 | 3.8 | 5.8 |
| Arjo | 1856 | 20 | 20 | - | 2565 | 1973-2004 | - | 8.3 | - |
| Debre Birhan | - | - | 6 | 12.7 | 2750 | | 1976-2004 | - | 7.73 |
| Dangla | - | - | 4 | 16.9 | 2140 | | 1986-2006 | - | 9.20 |
| Fiche | - | - | 4 | 14.1 | 2750 | | 1976-2004 | - | 10.2 |
| Mehal Meda | - | - | 4 | 12.5 | 3040 | | 1975-2004 | - | 10.8 |
| Assosa | - | - | 5 | 21.1 | 1600 | | 1965-2004 | - | 9.13 |
| Adet | - | - | 4 | 17.4 | 2080 | | 1986-2006 | - | 7.4 |

Table A3. Trend results of Mann-Kendall 'Z' statistic and 'p' value (in brackets) for stream flow time series. Bold numbers designate statistically significant trends at 5% level. Positive/negative sign of Z indicates increasing/decreasing trend, respectively.

| Streamflow variables | Catchment | | | | | | | | |
|---|---|---|---|---|---|---|---|---|---|
| | Gilgel Abay | Koga | Gumera | Rib | Jedeb | Chemoga | Muger | Guder | Neshi |
| Mean annual | -0.51 (0.60) | 1.46 (0.15) | 1.09 (0.28) | -0.09 (0.63) | -0.68 (0.50) | 0.15 (0.81) | 0.05 (0.96) | 0.87 (0.38) | **1.98 (0.05)** |
| Rainy season | -1.41 (0.14) | 0.85 (0.37) | **2.45 (0.01)** | -0.12 (0.89) | -0.74 (0.46) | -0.39 (0.63) | -0.45 (0.65) | 0.44 (0.61) | **2.08 (0.03)** |
| Dry season | 0.29 (0.76) | 1.21 (0.18) | 1.11 (0.27) | 0.00 (1.00) | -1.11 (0.27) | 0.22 (0.82) | 0.39 (0.70) | 1.32 (0.17) | 0.7 (0.40) |
| Small rainy season | **-2.65 (0.005)** | 0.67 (0.48) | **3.47 (0.00)** | -0.34 (0.71) | 0.25 (0.80) | 0.23 (0.80) | -0.71 (0.48) | **2.89 (0.01)** | 0.18 (0.82) |
| 1-day annual minima | **-4.49 (0.00)** | 0.64 (0.53) | 1.16 (0.24) | 0.66 (0.51) | 0.14 (0.89) | -1.70 (0.09) | **-2.01 (0.04)** | **2.59 (0.01)** | 0.02 (0.99) |
| 7-day annual minima | **-5.16 (0.00)** | 0.25 (0.8) | 1.13 (0.26) | 0.20 (0.84) | -1.19 (0.23) | -1.57 (0.12) | -1.78 (0.07) | **2.58 (0.01)** | 0.26 (0.80) |
| 1-day annual maxima | 0.06 (0.95) | 1.38 (0.17) | -0.87 (0.38) | -0.87 (0.38) | 1.37 (0.17) | 0.11 (0.91) | 0.73 (0.47) | **2.37 (0.02)** | **2.16 (0.03)** |
| 7-day annual maxima | -1.7 (0.09) | 0.28 (0.57) | 0.22 (0.45) | 0.76 (0.45) | -0.36 (0.72) | -1.23 (0.22) | 0.55 (0.58) | 0.43 (0.67) | **2.53 (0.01)** |

Table A4. Summarised results of Pettitt test for the mean annual and seasonal time scales. The first, second and third rows indicate for minimum, maximum and mean temperature, positive sign in the parentheses indicate increasing shift and negative sign indicate decreasing shift, numbers in bold designate p value for statistically significant change points.

| Station | Mean annual | Rainy season | Dry season | Small rainy season |
|---|---|---|---|---|
| | 2000 ( +), **0.01** | 1995 (+), **0.00** | 1999 (+), **0.04** | 1980 (+), **0.01** |
| | 2001 (+),**0.01** | 2002 (+), **0.03** | 1995 (+), 0.42 | 1998 (+), **0.01** |
| Adet | 2001 (+),**0.01** | 1995 (+), **0.00** | 1994 (+), 0.13 | 2001 (+), **0.01** |
| | 1986 (+), **0.00** | 1989 (+), **0.00** | 1987 (+), **0.04** | 1986 (+), **0.02** |
| Alem | 1984 (-), 0.21 | 1988 (-), 0.34 | 1999 (+), 0.19 | 1997 (+), 0.50 |
| Ketema | 1993 (+), **0.01** | 1995 (+), **0.04** | 1994 (+), 0.19 | 1991 (+), 0.08 |
| | 1988 (+), 0.66 | 1975 (+), 0.27 | 1987 (+), 0.45 | 2001 (+), 0.62 |
| | 1976 (+), 0.09 | 1978 (+), 0.22 | 1978 (+), 0.25 | 1976 (+), 0.27 |
| Assosa | 1981 (+), 0.19 | 1981 (+), 0.12 | 1982 (+), 0.28 | 1982 (+), 0.20 |
| | 1979 (+), **0.00** | 1979 (+), **0.00** | 1979 (+), **0.00** | 1979 (+), **0.01** |
| | 1992 (+), **0.00** | 1993 (+), **0.03** | 1991 (+), **0.00** | 1995 (+), **0.03** |
| Bahir Dar | 1979 (+), **0.01** | 1979 (+), **0.00** | 1983 (+), **0.00** | 1979 (+), **0.00** |
| | 1983 (+), **0.00** | 1990 (+), **0.00** | 1983 (+), **0.00** | 1983 (+), **0.00** |
| Debre | 1993 (+), **0.05** | 1992 (+), 0.41 | 1993 (+), 0.42 | 1993 (+), 0.35 |
| Markos | 1986 (+), **0.00** | 1990 (+), **0.00** | 1983(+), **0.00** | 1990 (+), **0.02** |
| | 1979 (+), **0.01** | 1986 (+), **0.00** | 1978 (+), 0.06 | 1982 (+), **0.01** |
| | 1992 (+), **0.03** | 1991 (+), **0.01** | 1989 (+), **0.00** | 1996 (+), 0.07 |
| Bedelle | 1982 (+), **0.02** | 1980 (+), **0.01** | 1982 (+), **0.02** | 1986 (+), **0.03** |
| | 2001 (+), **0.02** | 2001(+), 0.08 | 1992 (-), 0.41 | 2001 (+), 0.24 |
| | 2001 (+), **0.04** | 2002 (+), 0.07 | 1995 (+), 0.66 | 2002 (+), **0.05** |
| Dangla | 2001 (+), **0.03** | 2001 (+), **0.05** | 1999 (+), 0.62 | 2001 (+), 0.06 |
| | 1985 (+), 0.32 | 1987 (+), 0.23 | 1985 (+), 0.39 | 1983 (+), 0.44 |
| | 1994 (+), **0.03** | 1985 (-), 0.41 | 1995 (+), 0.11 | 1990 (+), **0.01** |
| Debre Birhan | 2001 (+), 0.16 | 1987 (+), 0.27 | 1985 (+), 0.35 | 1983 (+), **0.03** |
| | 1988 (+), 0.23 | 1988 (+), 0.05 | 1987 (+), 0.32 | 1988 (+), 0.07 |
| | 1983 (+), **0.05** | 1992 (-), 0.26 | 1989 (-), 0.58 | 1997 (+), 0.07 |
| | 1986 (+), 0.48 | 1992 (-), 0.67 | 1982 (-), 0.72 | 1988 (+), **0.01** |
| Fiche | | | | |
| | 1979 (-), 0.48 | 1995 (+), 0.11 | 1983 (+), 0.18 | 1987 (+), 0.11 |
| | 1987 (+), 0.41 | 1989 (+), 0.25 | 1987 (+), 0.47 | 1983 (+), 0.70 |
| Fincha | 2001 (+), 0.06 | 2000 (+), **0.05** | 1994 (+), **0.02** | 2002 (+), 0.72 |
| | 1981 (+), 0.16 | 1994 (+), 0.11 | 1981 (+), 0.37 | 1986 (+), 0.13 |
| | 1996 (+), **0.02** | 1996 (+), 0.33 | 1992 (+), 0.20 | 1996 (+), **0.01** |
| Mehal Meda | 1994 (+), 0.34 | 1994(+), 0.07 | 1982 (-), 0.34 | 1990 (+), **0.00** |
| | 1992 (-), **0.02** | 1991(-), 0.13 | 1992 (-), **0.00** | 1991 (-), 0.17 |
| | 1994 (+), **0.01** | 1995 (+), **0.03** | 1994 (+), 0.25 | 1997 (+), **0.03** |
| Nekemet | 1991 (-), 0.18 | 1994 (+), 0.38 | 1992 (-), 0.14 | 1998 (+), 0.05 |

Table A5. Summarized results of Pettitt test for the stream flows of nine gauging stations in the Abay/Upper Blue Nile basin. Numbers in the first row indicate the change point time, the positive sign in the parentheses indicate increasing shift and negative sign indicate decreasing shift, and the second row indicates 'p' value and numbers in bold designate statistically significant change of point at 5% level.

| Sreamflow variables (m³/s) | Catchment | | | | | | | | |
|---|---|---|---|---|---|---|---|---|---|
| | Gilgel Abay | Koga | Gumera | Rib | Jedeb | Chemoga | Muger | Guder | Neshi |
| Mean annual flow | 1981(-) (0.21) | **1995(+) (0.02)** | **1987(+) (0.01)** | 1986(+) (0.53) | **1983(-) (0.05)** | 1993(-) (0.82) | 1993(+) (0.7) | 1982(+) (0.23) | **1986(+) (0.00)** |
| Rainy season | 1981(-) (0.30) | 1995(+) (0.18) | **1987(+) (0.01)** | 1978(+) (0.76) | 1983, (-) (0.06) | 1993(-) (0.28) | 1994(-) (0.057) | 1982(+) (0.7) | **1986(+) (0.00)** |
| Dry season | 2000(+) (0.69) | 1989(+) (0.10) | 1991(+) (0.23) | 1981(+) (0.86) | 1993(-) (0.08) | 2000(+) (0.79) | 1977(-) (0.56) | 1995(+) (0.19) | 1995(+) (0.73) |
| Small rainy season | **1997(-) (0.01)** | 1986(+) (0.19) | **1991(+) (0.00)** | 1979(-) (0.19) | 1992(+) (0.41) | 1986(+) (0.5) | 1997(-) (0.09) | **1988(+) (0.01)** | **1984(+) (0.02)** |
| 1-day annual minima | **1994(-) (0.00)** | 1991(+) (0.13) | **1990(+) (0.00)** | 1988(+) (0.42) | 1992(+) (0.69) | 1993(-) (0.07) | **1997(-) (0.03)** | **1998(+) (0.00)** | 1985(+) (0.24) |
| 7-day annual minima | **1994(-) (0.00)** | 1988(+) (0.22) | **1990(+) (0.00)** | 1979(-) (0.49) | 1997(-) (0.23) | 1993(-) (0.11) | **1997(-) (0.02)** | **1998(+) (0.00)** | 1985(+) (0.06) |
| 1-day annual maxima | 1988(+) (0.40) | **1994(+) (0.01)** | 1997(-) (0.50) | **1981(-) (0.00)** | **2002(+) (0.01)** | 2001(+) (0.76) | 1988(+) (0.28) | **1988(+) (0.02)** | **1986(+) (0.00)** |
| 7-day annual maxima | 1997(-) (0.07) | 1994(+) (0.14) | 1987(+ (0.13) | **1980(-) (0.04)** | 1983(+) (0.26) | **1994(-) (0.03)** | 1988(+) (0.54) | 1982(+) (0.43) | **1986(+) (0.00)** |

## Appendix B

Figure 4.3 (see. Chapter 4) illustrates the model structure of the dynamic monthly water balance model. Rainfall $P(t)$ at time step $t$ partitions into direct runoff $Q_d(t)$ and $X(t)$. $X(t)$ is a lumped water balance component known as catchment rainfall retention which consists of the amount of retained water for catchment water storage $\dfrac{ds}{dt}$, $E(t)$ and recharge $R(t)$.

$$P(t) = Q_d(t) + X(t) \qquad\qquad (A4.1)$$

Where; $P(t)$, and $Q_d(t)$ are monthly rainfall and direct runoff, respectively. The units of all fluxes are mm/month.

Analogous to Budyko's hypotheses, Zhang et al. (2008) defined the demand limit for $X(t)$ to be the sum of dS/dt and potential evaporation $E_P$, which is termed as $X_0(t)$ and the supply limit as $P(t)$. If the sum of available storage capacity and potential evaporation is very large as compared to the supply; then $X(t)$ approaches $P(t)$. Whereas if the sum of available storage capacity and potential evaporation is smaller than the supply, $X(t)$ approaches $X_0(t)$. This postulate can be written as:

$$\frac{X(t)}{P(t)} \to 1, \quad \frac{X_0(t)}{P(t)} \to \infty, \text{ for very dry conditions} \qquad\qquad (A4.2)$$

$$X(t) \to X_0(t), \text{ as } \frac{X_0(t)}{P(t)} \to 0, \text{ for very wet conditions} \qquad\qquad (A4.3)$$

The catchment rainfall retention can be expressed as:

$$X(t) = \begin{cases} P(t)F\left[\dfrac{X_0(t)}{P(t)}, \alpha_1\right], & P(t) \neq 0 \\ 0, & P(t) = 0 \end{cases} \qquad\qquad (A4.4)$$

Where F [-] is the Fu-curve equation (4.4) described in chapter 4, and $\alpha_1$ is the retention efficiency, whereby larger $\alpha_1$ values result in more rainfall retention and less direct runoff.

From equation (A4.1) and equation (A4.4) the direct runoff is calculated as:

$$Q_d = P(t) - X(t) \qquad\qquad (A4.5)$$

The water availability $W(t)$ for partitioning can be computed as:

$$W(t) = E(t) + S(t) + R(t) \qquad\qquad (A4.6)$$

Sankarasubramania and Vogel (2003) defined the evaporation opportunity as maximum water that can leave the basin as evaporation at any given time $t$ using equation A4.7.

$$Y(t) = E(t) + S(t) \qquad\qquad (A4.7)$$

For the partition of available water, the demand limit for $Y_0(t)$ is the sum of the soil water storage capacity $(S_{max})$ and potential evaporation $E_P(t)$ termed as $Y_0(t)$, while the supply limit is the available water $W(t)$. Analogous to Budyko's hypothesis, Zhang *et al.* (2008) postulated that:

$$\frac{Y(t)}{W(t)} \to 1, \quad \text{as} \quad \frac{Y_0(t)}{W(t)} \to \infty, \text{ for dry conditions, and} \tag{A4.8}$$

$$Y(t) \to Y_0(t), \quad \text{as} \quad \frac{Y_0(t)}{W(t)} \to 0, \text{ for wet conditions.} \tag{A4.9}$$

The evaporation opportunity can be computed as:

$$Y(t) = \begin{cases} W(t)F\left[\dfrac{E_P(t)+S_{max}}{W(t)}, \alpha_2\right], & W(t) \neq 0 \\ 0, & W(t) = 0 \end{cases} \tag{A4.10}$$

From equations (A4.6) and (A4.7) the groundwater recharge $R(t)$ can be estimated as:

$$R(t) = W(t) - Y(t) \tag{A4.11}$$

To estimate the evaporation, the demand limit is $E_P(t)$ and the supply limit is $W(t)$, then $E(t)$ can be computed as:

$$E(t) = \begin{cases} W(t)F\left[\dfrac{E_P(t)}{W(t)}, \alpha_2\right], & W(t) \neq 0 \\ 0, & W(t) = 0 \end{cases} \tag{A4.12}$$

Where $\alpha_2$ is a model parameter representing the evaporation efficiency. Equation (A4.12) is similar with equation (4.4) with the exception that precipitation is replaced by water availability $W(t)$ to take into account the effect of catchment water storage.

Zhang et al. (2008) also emphasized that equation (A4.10) and (A4.12) use the same parameter. This is due to the fact that the groundwater recharge is essentially determined from evaporation efficiency. As evaporation efficiency becomes larger, i.e. for large values of $\alpha_2$, the recharge is diminished. Using equation (A4.10 and A4.12), the soil water storage can be computed as:

$$S(t) = Y(t) - E(t) \tag{A4.13}$$

The groundwater storage is treated as linear reservoir. Thus, the base flow $Q_b(t)$ and the groundwater storage $G(t)$ can be modelled as:

$$Q_b(t) = dG(t-1) \tag{A4.14}$$

$$G(t) = (1 - d\Delta t)G(t-1) + R\Delta t \tag{A4.15}$$

Where; $Q_b$, $G$, and $d$ are the baseflow [mm/month], groundwater storage [mm], and reservoir constant [1/month], respectively.

# SAMENVATTING

De watervoorraden in het Abay / Boven Blauwe Nijl stroomgebied zijn de levensbron van honderden miljoenen mensen die in dit gebied en verder stroomafwaarts leven. Het zorgt voor meer dan 60% van het totale water in de Nijl. Intensieve landbouw in ongeschikte grond en op hellingen, overbegrazing en gronderosie behoren tot de grootste problemen in het gebied. Landdegradatie heeft ook invloed op de hydrologie in het stroomgebied en op de beschikbaarheid van de watervoorraden. Vandaar dat duurzaam waterbeheer in het stroomgebied nodig is, wat diepgaand begrip van de hydrologie in het gebied vereist. Dit zou mogelijk bereikt kunnen worden met behulp van de beoordeling van hydrologische variabiliteit, het onderzoeken van de wederzijdse interactie tussen landgebruik op de hydrologische reacties, en een gedetailleerd inzicht in de neerslag-runoff processen.

Hoewel gedetailleerde kennis van de hydrologie van het stroomgebied belangrijk is vanuit zowel wetenschappelijk als operationeel oogpunt, wordt dit belemmerd door de schaarse hydro-meteorologische gegevens. Daarnaast zijn de variaties in ruimte en tijd van neerslag en andere meteorologische parameters, alsmede van fysiografische kenmerken groot, met als gevolg dat de hydrologische processen in het stroomgebied vrij complex zijn. Bovendien zijn deze processen zelden eerder onderzocht. Daarom heeft het analyseren van de hydrologische processen op verschillende ruimtelijke en temporele schalen de hoogste prioriteit om het effect van veranderingen in het stroomgebied te kunnen voorspellen en om duurzaam waterbeheer te kunnen begeleiden. Dit proefschrift richt zich op het karakteriseren en kwantificeren van processen in het stroomgebied en het modelleren in het gebied door intensieve veldmetingen en een set van verschillende modelbenaderingen die elkaar aanvullen in de reeks ruimte- en tijdschalen. Verschillende methoden, waaronder de lange termijn trendanalyse, het verzamelen van veldgegevens en gecombineerde stabiele isotopen en procesmatige neerslag-runoff modellen zijn uitgevoerd.

De lange-termijn trends (1954-2010) van neerslag, temperatuur en afvoerstroming werden geanalyseerd op drie meetstations in het stroomgebied. Mann-Kendall en Pettitt tests zijn gebruikt voor respectievelijk de trend en de detectie van knikpunt analyse. De resultaten toonden abrupte veranderingen en gerelateerde opwaartse en neerwaartse veranderingen in respectievelijk tijdreeksen van temperatuur en afvoermetingen. Echter, tijdreeksen van neerslag leverden geen statistisch significante trends met een significantie niveau van 5% van de gemiddelde jaarlijkse en seizoensgebonden schalen bij de onderzochte stations op. Stijgende trends in de temperatuur bij verschillende weerstations voor de gemiddeld jaarlijkse, regenachtige, droge en korte regenseizoenen werden duidelijk. Bijvoorbeeld: de gemiddelde temperatuur op Bahir Dar in het Lake Tana sub-stroomgebied is toegenomen met een snelheid van 0.5, 0.3 en 0.6°C / decennium voor respectievelijk het belangrijkste regenseizoen (juni tot september), het korte regenseizoen (maart tot mei) en het droge seizoen (oktober-februari).

Om de grootschalige hydrologische dynamiek van het stroomgebied te begrijpen, zijn waterbalansen berekend voor twintig geselecteerde stroomgebieden, variërend van 200 tot 173.686 km². Een eenvoudig model op basis van de Budyko hypothese naar aanleiding van een top-down modelmatige benadering is gebruikt om de waterbalans op de jaarlijkse en maandelijkse tijdschalen te analyseren. De resultaten tonen aan dat op basis van een jaarlijkse tijdschaal alle stroomgebieden niet door dezelfde Budyko curve gerepresenteerd kunnen worden. Tussen-jaarlijkse variabiliteit van neerslag, landgebruik / bodembedekking, bodemtype, geologie en topografie zijn hoogstwaarschijnlijk de redenen voor de verschillende reacties van een stroomgebied. De resultaten van het jaarlijkse waterbalansmodel werden verbeterd door het verkleinen van de tijdschaal in maandelijkse schalen en door het toepassen van de maandelijkse dynamiek van de opslag van bodemvocht in de wortelzone in het model. Het maandelijkse model gaf betere kalibratie- en validatie-resultaten van de afvoerstromingsdynamiek voor de meerderheid van de stroomgebieden.

De resultaten van de trendanalyses en waterbalansberekeningen tonen aanzienlijke verschillen in stromingsregimes en hydrologische reacties verspreid over de Abay / Boven Blauwe Nijl stroomgebieden. Om meer inzicht te krijgen in de redenen van deze variabiliteit is er een diepgaand onderzoek uitgevoerd naar de hydrologische reacties op veranderingen in het landgebruik / bodembedekking voor het meso-schaal Jedeb stroomgebied (296 km²). Het Jedeb stroomgebied wordt gekenmerkt door intensieve landbouw en een aanzienlijke uitbreiding van het landbouwareaal in de afgelopen decennia, bijvoorbeeld: landbouwgrond is in de afgelopen 50 jaar met 17% toegenomen.

Met behulp van statistische tests voor dagelijkse gegevens en maandelijkse gemodelleerde gegevens gebaseerd op de Budyko benadering is het effect van veranderingen in landgebruik op de afvoerstroming gekwantificeerd voor het Jedeb stroomgebied. De resultaten tonen een significante verandering van de dagelijkse stromingskenmerken waargenomen tussen 1973 en 2010. De piekafvoer wordt verhoogd, in andere woorden, de reactie is schokkender geworden. Er is een aanzienlijke toename in de toe- en afnamegraad van de hydrograph alsook in het aantal lage stromingspulsen. De afvoerpulsen laten een daling in de tijd zien, wat een teken is van toegenomen 'schokkerigheid' van het stroomgebied. De resultaten van het Budyko model toonden een verandering van modelparameters in de loop van de tijd, die ook kan worden verklaard door een verandering in landgebruik / landbedekking. De model parameters die bodemvochtigheid aangeven, gaven een geleidelijk dalende trend aan, wat een vermindering van opslagcapaciteit impliceert, die kan worden verklaard door de vergroting van de landbouw in het stroomgebied met daarbij een relatieve afname van natuurlijke gronden, zoals bos. De resultaten van de maandelijkse afvoerduurlijn analyse gaven grote veranderingen van het stromingsregiem aan in de loop van de tijd. De grote afvoerstromingen zijn tussen 1990 en 2000 met 45% gestegen. Dit in tegenstelling tot kleine afvoerstromingen, die tussen 1970 en 2000 met 85% daalden. Deze resultaten zijn relevant voor duurzaam waterbeheer in het Jedeb stroomgebied en in andere vergelijkbare stroomgebieden in de omgeving.

De karakterisering van stabiele milieu-isotopen om de gemiddelde verblijftijd en de runoffcomponenten in de belangrijkste stroomgebieden te identificeren heeft nuttige informatie voor het beheer van het stroomgebied opgeleverd. Zowel in Jedeb als in het naburige meso-schaal stroomgebied Chemoga, zijn in-situ isotopen monsters van neerslag, bronwater en afvoerstroming verzameld en geanalyseerd. De resultaten tonen aan dat de isotopensamenstelling van neerslaggebieden gemarkeerd wordt seizoensgebonden variaties, wat kan duiden op verschillende vochtbronnen van de neerslag. De Atlantische Oceaan, het Congo stroomgebied, de Boven Witte Nijl en de Sudd moerassen zijn geïdentificeerd als potentiële brongebieden van vocht tijdens het belangrijkste regenachtige (zomer)seizoen, terwijl de Indisch-Arabische, en de Middellandse Zee de belangrijkste brongebieden van vocht zijn tijdens het korte regenseizoen en de droge (winter)seizoenen. Resultaten uit de hydrograph met een seizoensgebonden tijdschaal geven de dominantie weer van het water met een gemiddelde van 71% en 64% van de totale runoff tijdens het natte seizoen in respectievelijk de Chemoga en Jedeb stroomgebieden. De resultaten tonen verder aan dat de gemiddelde verblijftijden van het water 4,1 en 6,0 maanden bedragen voor respectievelijk de gehele Chemoga en Jedeb stroomgebieden.

Tenslotte is er op basis van de veldmetingen, reactietijden van stroomgebieden en waterbalans modellen, een gedetailleerd procesmatig hydrologisch model ontwikkeld met PCRaster software. Het model werd ontwikkeld om de hydrologische processen van de twee meso-schaal stroomgebieden Chemoga en Jedeb op een dagelijkse tijdschaal te bestuderen. De bewerkte meetgegevens, dagelijkse afvoerreeksen en verschillende ruimtelijke gegevens werden gebruikt om het model te ontwikkelen op een raster van $200 * 200 \ m^2$. Drie verschillende weergaves van het model zijn toegepast om een geschikte modelstructuur te verkrijgen. Kalibratie van het model en evaluatie van onzekerheden zijn geïmplementeerd binnen het 'Generalized Likelihood Uncertainty Estimation' (GLUE) kader. Tijdens de kalibratie van het model zijn parameters geconditioneerd met behulp van alleen de afvoergegevens en informatie over stabiele isotopen in de vorm van de fractie van nieuw en oud water. De resultaten van de verschillende weergaves van het model zijn geëvalueerd in termen van prestatie-indicatoren, parameter identificeerbaarheid en beperkt voorspelbare onzekerheid.

De modelresultaten hebben duidelijk aangetoond dat parameters beter herkenbaar bleken te zijn en een beperkt voorspelbare modelonzekerheid bleken te hebben bij het gebruik van stabiele isotopeninformatie als aanvulling op afvoermetingen. De stabiele isotopen verstrekten aanvullende informatie over stroompaden en runoffcomponenten in de twee stroomgebieden en ondersteunen zodoende het proces-gebaseerde modeleren. Het is gebleken dat de verzadigde overtollige bovengrondse afvoerstroming hoogstwaarschijnlijk het dominante runoff genererende proces is tijdens regenval / runoff gebeurtenissen in beide stroomgebieden, wat in verwachting is met de veldwaarnemingen. De modelonderzoeken hebben aangetoond dat de twee stroomgebieden niet even goed kunnen worden gemodelleerd met dezelfde modelstructuur. Dit wordt veroorzaakt door verschillen in neerslag-runoff processen die veroorzaakt zijn door de verschillende grootte van de watergebieden per stroomgebied. Daarom wordt geconcludeerd dat één enkele modelstructuur voor de gehele Abay / Boven Blauwe Nijl niet alle dominante hydrologische processen van de sub-stroomgebieden kan weergeven. Vandaar dat semi-gedistribueerde modellen met gedistribueerde inputgegevens essentieel zijn, waarbij individuele runoff-processen van de verschillende landschapselementen (wetlands, heuvels en plateau) meegenomen worden als inputgegevens.

Gezien de resultaten verkregen met gedetailleerde hydrologische metingen bij de twee meso-schaal sub-stroomgebieden, de hydrologische reacties op veranderingen van het landgebruik / landbedekking, de lange termijn trend analyse van hydro-meteorologische parameters, het grootschalige Budyko modelleren, en tenslotte het gedetailleerd conceptueel verdeeld modelleren, heeft deze studie gezorgd voor diepgaande inzichten en een beter begrip van de hydrologische processen binnen het Boven Blauwe Nijl stroomgebied. Dit is belangrijk voor zowel het beheer en de duurzame ontwikkeling van de hulpbronnen van de Blauwe Nijl als voor toekomstig onderzoek in het stroomgebied.

# About the author

Sirak Tekleab was born in Addis Ababa, Ethiopia, on September, 24th 1975.
He completed his first degree in Irrigation Engineering at Arba Minch
Water Technology Institute (AWTI), Ethiopia, in 2001. Then, he worked for 3 years at the South
Irrigation Development Authority Ethiopia until November 2004. In the organization, he was
actively involved in planning, designing and implementing water resources development projects
in the region. Furthermore, in 2003 he participated as water resources engineer in the re-settlement
program, which is one of the development programs, the country's endeavour. After having three
years of practical work experience, he has joined Hawassa University, Institute of Technology, as
a graduate assistance in November 2004. He worked for two years at the University offering
various water related regular courses for undergraduate students. Furthermore, he was also actively
involved in the Engineering panel section of the University, serving the community by designing
and supervising various civil and water works.

After five years of work experience Sirak joined Arba Minch University for his post-graduate
studies in September 2006. He got his Master of Science degree in Hydrology and Water
Resources Management in September 2008. From 2008 to 2014, he has worked as PhD research
fellow at UNESCO-IHE/TU-Delft, The Netherlands. During the course of his PhD, he studied the
Abay/Upper Blue Nile basin hydrology at various spatio-temporal scales using a combination of
different methods. Moreover, he has supervised MSc students and served as a reviewer for
different peer-reviewed international journals. Further more, he attended the course requirement
for the graduate school for Socio-economic and Natural Sciences of the Environment (SENSE).
Recently, in Sept. 2013 Sirak founded the Water Resources Development Consulting Firm, a
category V accredited firm by the Ethiopian Ministry of Water, Irrigation, and Energy. This will
enable him to contribute even more to the water sector in the future.

## Peer-reviewed publications:

Tekleab, S., Uhlenbrook, S., Savenije, H.H.G., Mohamed, Y., and Wenninger, J. (2014). Catchment
Modelling through the use of stable environmental isotopes and field observations in the
Chemoga and Jedeb meso-scale catchments. (Under review: Hydrological Science Journal).

Tekleab, S., Wenninger, J., and Uhlenbrook, S. (2014). Characterisation of stable isotopes to identify
residence times and runoff components in two meso-scale catchments, Abay/Upper Blue Nile
basin, Ethiopia. Hydrol. Earth Syst. Sci., 18, 2415–2431, doi: 10.5194/hess-18-2415.

Tekleab, S., Mohamed, Y., Uhlenbrook, S., and Wenninger, J. (2014). Hydrologic responses to land
cover change: the case of Jedeb meso-scale catchment, Abay/Upper Blue Nile basin, Ethiopia.
Hydrol. Process. J., doi: 10.1002/hyp.9998, 28, 5149-5161.

Tekleab, S., Mohamed, Y, and Uhlenbrook, S. (2013). Hydro-climatic trends in the Abay/Upper Blue Nile basin, Ethiopia. Journal of Physics and Chemistry of the Earth. Doi: 10.1016/j.pce. 2013.04.017.

Tekleab, S., Uhlenbrook, S., Mohamed, Y., Savenije, H.H.G., Temesgen, M, and Wenninger, J. (2011). Water balance modelling of the upper Blue Nile catchments using a top-down approach. Hydrology and Earth System Science 15: 2179–2193 doi: 10.5194/hess-15-2179.

## Conference papers:

Tekleab, S., Mohamed, Y., Uhlenbrook, S., and Wenninger, J. (2013). Hydrologic responses to land cover change: the case of Jedeb meso-scale catchment, Abay/Upper Blue Nile basin, Ethiopia. Oral presentation at International conference on New Nile perspective, May 6-8 2013. Khartoum, Sudan.

Koch, F.J., van Griensven A., Uhlenbrook, S., Tekleab, S., Teferi, E. (2012). The Effects of Land use Change on Hydrological Responses in the Choke Mountain Range (Ethiopia) – A new Approach Addressing Land Use Dynamics in the Model SWAT. 6th International Congress on Environmental Modelling and Software (iEMSs), 1-5 July 2012, Leipzig, Germany (oral presentation).

Tekleab, S., Wenninger, J., and Uhlenbrook, S. (2012).Characterization of stable isotope in the Choke Mountains range, Abay/Upper Blue Nile basin. Poster presentation at International association for hydrological science; PUB symposium, Delft, The Netherlands.

Tekleab, S., Uhlenbrook, S., Mohammed, Y., Savenije, H.H.G., Temesgen, M., and Wenninger, J. (2011). Water balance modelling of Chemoga and Jedeb watershed; headwater tributaries of Abay/Blue Nile basin, Ethiopia (Choke Mountain development research project proceeding).

Tekleab, S., and Ayalew, S. (2009). Watershed Modelling of Lake Tana sub-basin using SWAT Model. Oral presentation at Nile Basin first annual research conference, Dar es Salaam, Tanzania.

Printed and bound by CPI Group (UK) Ltd, Croydon, CR0 4YY

21/10/2024

01777101-0007